VISIONS OF A DIGITAL NATION

History of Computing
William Aspray and Thomas J. Misa, editors

A complete list of the titles in this series appears in the back of this book.

VISIONS OF A DIGITAL NATION

MARKET AND MONOPOLY IN BRITISH TELECOMMUNICATIONS

JACOB WARD

The MIT Press
Cambridge, Massachusetts
London, England

The MIT Press would like to thank the anonymous peer reviewers who provided comments on drafts of this book. The generous work of academic experts is essential for establishing the authority and quality of our publications. We acknowledge with gratitude the contributions of these otherwise uncredited readers.

This book was set in Stone Serif and Stone Sans by Westchester Publishing Services. Printed and bound in the United States of America.

The "T circle logo" and "stylised Telecom logo" shown within the cover image are trade marks of British Telecommunications Plc.

Library of Congress Cataloging-in-Publication Data

Names: Ward, Jacob (Science and technology historian), author.
Title: Visions of a digital nation : market and monopoly in British
    telecommunications / Jacob Ward.
Description: Cambridge, Massachusetts : The MIT Press, [2023] |
    Series: History of computing | Includes bibliographical references
    and index.
Identifiers: LCCN 2023013902 (print) | LCCN 2023013903 (ebook) |
    ISBN 9780262546294 (paperback) | ISBN 9780262375535 (epub) |
    ISBN 9780262375528 (pdf)
Subjects: LCSH: Telecommunication—Great Britain—History—20th century. |
    British Telecom. | Digital communications—Economic aspects—Great
    Britain. | Digital communications—Political aspects—Great Britain. |
    Privatization—Great Britain. | Neoliberalism—Great Britain.
Classification: LCC HE8094 .W37 2023 (print) | LCC HE8094 (ebook) |
    DDC 384.0941—dc23/eng/20230724
LC record available at https://lccn.loc.gov/2023013902
LC ebook record available at https://lccn.loc.gov/2023013903

10 9 8 7 6 5 4 3 2 1

CONTENTS

ACKNOWLEDGMENTS

I couldn't have written this book without the help and support of many friends, family, and colleagues along the way. This book started as a PhD project at University College London's Department of Science and Technology Studies, where I was supervised by Jon Agar. I don't think words can do justice to how much I and this book benefited from Jon's patience, insight, and feedback. At UCL, Tiago Mata, my internal examiner, gave great feedback and career advice, and Joe Cain helped me think through the very first version of the book proposal. Graeme Gooday, from Leeds, was my external examiner and gave helpful advice on how to move forward from the PhD. The PhD project was also a collaborative project with the Science Museum, London, and so, from the Science Museum, I'd especially like to thank Tilly Blyth, Tim Boon, and John Liffen for their support and advice along the way. This project was also part of a broader project on the history of Post Office and BT R&D, and I feel very fortunate that I could do that project alongside Alice Haigh and Rachel Boon, the two other PhD candidates on the project. The project was funded by the UK's Arts and Humanities Research Council, which also deserves thanks as, without that support, I couldn't have done the PhD in the first place.

Since leaving UCL, I've worked at the University of Oxford and Maastricht University. From Oxford, I must especially thank Ursula Martin. The year I spent as a postdoctoral researcher for Ursula gave me the space and security that I needed to rethink this research and begin turning it into something worthy of a book. Chris Hollings also provided great support and advice along the way, as did Alexandra Franklin at the Bodleian, who gave me my first postdoctoral position, the Byrne-Bussey Marconi Fellowship. After Oxford, I joined the History Department and STS Research Programme (MUSTS) at Maastricht University, which has been a fantastic new home. From Maastricht, I'd especially like to thank Cyrus Mody and Vincent Lagendijk, both of whom read through the entire manuscript and gave really useful feedback that has elevated the book. I'd also like to thank Sally Wyatt, who led a discussion on chapter 4 at the MUSTS annual research day, and all those who attended, read, and commented. Finally, Anique Hommels has been a great mentor since I joined Maastricht and gave very useful advice on an early version of the book proposal.

This book would not have been possible without the people working at the archives and museums that I visited. The staff at BT Archives, who gave me such privileged and extensive access, were fundamental to the project, especially David Hay, James Elder, and Anne Archer. I also spent several months as a Lemelson Fellow at the National Museum of American History. From there, and also from the National Air and Space Museum, I was given great advice and made to feel very welcome by Eric Hintz, Arthur Daemmrich, Teasel Muir-Harmony, and Martin Collins. I also must thank Thomas and Mary Edsall, who were wonderful landlords during my stay, and Alana Staiti, who invited me to Cornell for a really useful opportunity to present my research to the STS department. Finally, I'd like to thank all the staff from across the archives that I used during this project: the British Postal Museum and Archive, the National Archives, the Bank of England Archive, the National Museum of American History Archives Center (especially Kay Peterson, who helped me secure permissions for some images in chapter 6), the National Air and Space Museum Archives Center, and the George Washington University Special Collections Center.

This book has also benefited greatly from the opportunities that I've had to publish parts of this research along the way. Earlier writing based on parts of chapter 2 appears in "Nineteen Eighty-Four in the British Telephone System," in Jenny Andersson and Sandra Kemp's edited collection, *Futures* (Oxford University Press, 2021)—and I'd like to especially thank Jenny for kick-starting my interest in the history of futurology, which continues beyond this book. Parts of chapters 2 and 3 also appear in "Computer Models and Thatcherist Futures," in *Technology & Culture* (2020), and a very early version of chapter 6 appears in "Oceanscapes and Spacescapes in North Atlantic Communications," in *Histories of Technology, the Environment and*

Modern Britain (UCL Press, 2018), which I coedited with Jon Agar. Finally, an earlier version of chapter 7 appears in "Financing the Information Age," in *Twentieth Century British History* (2019), and I'd like to especially thank the anonymous reviewers for showing me how my research could speak beyond the history of science and technology.

The MIT Press gave this book life, and I'd especially like to thank Tom Misa and Katie Helke for steering this project: Tom, for his kind words and for helping process the reviewers' feedback; Katie, for her advice throughout the years, from when I was still finishing the PhD project, on how to go about the whole process of writing a scholarly book. I'd also like to thank Laura Keeler and Suraiya Jetha, for advising on many details throughout, and the reviewers, including Valérie Schafer, who really helped me think about what matters, for both the proposal and the manuscript.

Finally, I couldn't have spent so much time and energy on this book if I didn't have my friends and family to lean on. My parents, Vanessa and Michael, have given so much support, and my wife, Louisa, has always been there for me, through both the highs and the lows. This book is dedicated to them.

ABBREVIATIONS

ACE	Automatic Computing Engine
AGSD	Advisory Group on Systems Definition
ALEM	A Local Exchange Model
AT&T	American Telephone & Telegraph
BoE	Bank of England
BPSD	Business Planning and Strategy Department
BT	British Telecom
CBI	Confederation of British Industry
CCITT	Consultative Committee for International Telegraphy and Telephony
CIE	Committee on Invisible Exports
COMSAT	Communications Satellite Corporation
CPRS	Central Policy Review Staff
CTC	City Telecommunications (Sub-)Committee
CTNE	Compañía Telefónica Nacional de España
DTI	Department of Trade and Industry
FCC	Federal Communications Commission
GEC	General Electric Company
GPO	General Post Office
GRACE	Group Routing and Charging Equipment
ICDM	Integrated Communications Demand Model
IEE	Institute of Electrical Engineers
INTELSAT	International Telecommunications Satellite Organization
IRC	Industrial Reorganisation Corporation

ISDN	integrated services digital network
ISO	International Organization for Standardization
ITAP	Information Technology Advisory Panel
ITT	International Telephone & Telegraph
ITU	International Telecommunications Union
LRPD	Long Range Planning Department/Division
LRPM	long-range planning model
MPBW	Ministry of Public Buildings & Works
MPT	Ministry of Posts & Telecommunications
NEDC	National Economic Development Council
NPL	National Physical Laboratory
OSI	Open Systems Interconnection
O&M	Treasury Organisation & Methods
POEU	Post Office Engineering Union
PSBR	public-sector borrowing requirement
PTT	Postal, Telegraph, and Telephone Service
SCU	strategic control unit
STC	Standard Telephones and Cables
STD	subscriber trunk dialing
STL	Standard Telecommunications Laboratories
TAT	transatlantic telephone/communications cable
UKTTF	United Kingdom Trunk Task Force
UPW	Union of Postal Workers

INTRODUCTION

On Friday, November 16, 1984, the world witnessed what was, at that time, the largest stock flotation in history. Margaret Thatcher's Conservative government had privatized the UK's national telecommunications provider, British Telecom, for approximately £3.6 billion.[1] BT's sale was not just any privatization. It was *the* privatization, paving the way throughout the world for privatization of not just digital infrastructure, but of infrastructure in general. BT's sale turned privatization into a core policy not only for the Thatcher government but also for other emerging neoliberal governments, which aimed to shrink the state and create new markets by selling state assets to the private sector.[2] Margaret Thatcher called privatization "one of Britain's most successful exports," boasting that it had trebled Britain's number of individual shareholders and "put a stop to the idea that inefficient management would always be subsidized by the taxpayers."[3] Kenneth Baker, Thatcher's minister for information technology, later wrote that BT's privatization "made possible all other public utility sales."[4] Thirty-five years after BT's privatization, Jeremy Corbyn's 2019 Labour Party manifesto vowed to reverse this change, renationalizing British Telecom and delivering free full-fiber broadband to all by 2030. This plan, however, was treated as shockingly radical. Boris Johnson called it a "crazed communist scheme," while Neil McRae, BT's chief network architect, tweeted that "labour [sic] plans broadband communism!"[5] Rejecting public broadband ownership naturalized the idea of private digital infrastructure, a legacy of Margaret Thatcher's government, while making public ownership appear alien. BT's privatization thus illustrates how private infrastructure ownership, one of the key features of neoliberalism, took shape.

But BT's sale was more than just a proof-of-principle for neoliberal economic policy. It also pushed the idea that digital infrastructure is best done by the private sector. When Patrick Jenkin, Thatcher's secretary of state for industry, announced the plan to sell BT in 1982, he told the British Parliament that "we need to free BT from traditional forms of government control" and linked this to digitalization, asserting that "British Telecommunications is already leading the information technology revolution in the United Kingdom. It could become a major world force."[6] That same year—which the government had christened Information Technology Year, also known as IT-82—Margaret Thatcher gave the keynote at an information technology conference at the Barbican Centre, London. Thatcher proclaimed that information technology required "free enterprise," and that this would soon include BT.[7] Kenneth Baker also gave a speech on the "information economy," arguing that privatization would stop the totalitarian "electronic state" from abusing information technology.[8] These remarks responded to the digitalization of Britain's telecom infrastructure. Since the late 1950s, scientists, engineers, and managers at the British Post Office, which ran the nation's telecommunications service until 1981, had been researching, planning, and building a new digital infrastructure for Britain. When James Merriman, the senior director of engineering at the Post Office, had announced his plans for digitalization in 1967, he proposed a public digital infrastructure that would allow every citizen to receive voice, video, and data communications to their home.[9] BT's sale, however, meant that Britain's digital infrastructure would look very different from Merriman's original plans.

This book recovers these lost horizons of nationalized digital infrastructure, showing engineers' and managers' plans for digitalization and how these plans collided with privatization. Communications infrastructure will always involve a mix of public and private ownership, with the shares of public and private in that mix constantly changing. In recent years, public and private advocates have pushed their visions of faster, fairer coverage for all. In the private sector, both Elon Musk's SpaceX and Jeff Bezos's Amazon are building global satellite internet infrastructures. On the public side, more national, earthly solutions hold sway. Since 2009, Australia has been building a national broadband network, touted as the largest infrastructure project in Australian history.[10] Bernie Sanders's Green New Deal pledged $150 billion to build public broadband networks across the US. This question of public or private ownership, however, is not just about the future but also about the past and present. Private US telecom monopolies have already raised prices and furthered inequality, while national laws governing the internet, such as the Great Chinese Firewall, reveal that the internet is not the private, borderless world that it seems.[11] This book is about a key moment in this history of public and private digital infrastructure, the moment when communications infrastructure became both significantly more private and significantly more digital.

This history of the digitalization and privatization of Britain's telecommunications infrastructure offers a way to understand how these changes were connected. Digitalization and privatization now seem two sides of the same coin. But this appearance is a historical construction, and neither digitalization nor privatization were things that merely happened to Britain's telecom infrastructure, or any other communications infrastructure. Digitalization began decades before privatization, and so, in the UK, was at first shaped by national ownership, only later turning from monopoly to market. These were changes that telecom engineers and managers made happen. In doing so, these engineers and managers not only changed how they understood digitalization, nationalization, and privatization but also changed how politicians, policymakers, and the public understood the relationships among digitalization, regulation, and infrastructure ownership. This book takes these engineers and managers as the central characters in this history and looks at how, starting in the 1950s, they began to think, plan, and enact the digitalization of British communications. Then, from the late 1970s, they began to alter course as they anticipated the end of their monopoly and the eventual privatization of their network. This book shows how, in doing so, monopoly and national ownership influenced digitalization, and vice versa, and how these associations changed with liberalization and privatization. The book argues that these changes in technology, ownership, and regulation cannot be understood in isolation, but must be understood together. First, however, this means understanding the wider histories of privatization and digitalization, and thus the relationship between technology and politics.

PRIVATIZATION: PRAGMATISM, IDEOLOGY, AND TECHNOLOGY

In the *longue durée* history of European infrastructure, regulatory and ownership changes seem to have been chiefly pragmatic. In the nineteenth century, neither municipal socialism nor anticapitalism played a great role in local and national governments taking over infrastructure.[12] Instead, state and municipal ownership in the nineteenth and early twentieth centuries across Europe met various political goals, such as mitigating private-sector cartelization and other market failures, or geopolitical and security concerns about information and resource flows. In the late twentieth century, privatization had its ideologues, such as the Austrian American management theorist Peter Drucker, who advocated privatization to reduce "government overload," and in members of the US New Right, who believed that it would create a property-owning democracy.[13] But for Margaret Thatcher's Conservative Party, elected in 1979, privatization was a way to balance the books after Britain's 1976 bailout by the International Monetary Fund, which imposed limits on how much money Britain's nationalized industries could borrow.[14] The British Treasury's shift to monetarist economics, aiming to lower inflation by limiting borrowing and reducing the budget deficit, further incentivized privatization.[15]

Privatization under Thatcher first took shape as sales of smaller government assets, like council houses, oilfield assets, and public-sector land, although council house sales were still quite radical.[16] The drive to find larger assets to sell meant that, by 1981, the Conservatives began to consider privatizing the state utility monopolies.[17] At this point, privatization's main goals were reducing the budget deficit by selling public assets and freeing state-owned enterprises from external financing limits so that they could borrow and invest. In October 1982, the Central Policy Review Staff, the Cabinet's "think tank," recommended bringing privatization to this next level, and, two years later, the government sold British Telecom. The government, however, did not present BT's sale as purely pragmatic but used it to promote individual choice and freedom over state ownership. The government explicitly targeted the British public as potential BT shareholders through an advertising campaign that included slogans such as, "You can share in BT's future" and "A public service goes public."[18] By early September 1984, the BT Share Information Office had received more than 300,000 requests for more information.[19] Between Friday, November 16, and Wednesday, November 28, 1984, when share applications for BT opened, underwriters in the City of London received two million applications from the British public, more applications than the number of individual shareholders in the entire nation before BT's sale. Recalling this popular enthusiasm, Nigel Lawson, then chancellor of the Exchequer, later heralded BT's sale as "the birth of people's capitalism."[20]

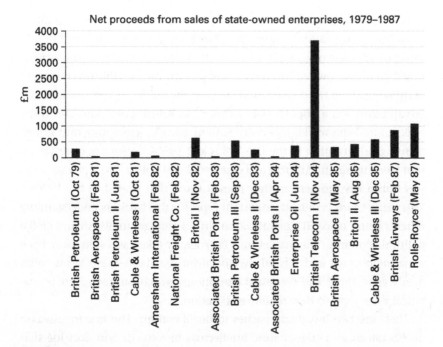

FIGURE 0.1
The proceeds from BT's privatization far exceeded those from any of the privatizations preceding it, as well as all of those in the three remaining years of the second Thatcher government. Source: Data taken from Parker, *The Official History of Privatisation: Volume 1* (2005).

To see privatization, particularly BT's privatization, as purely pragmatic thus misses its radical influence on the Thatcher government and Thatcherism. BT's unexpectedly successful share offer (figure 0.1) turned privatization from an economic to an ideological project, emphasizing "popular capitalism" and "share-owning democracy."[21] Wider share-ownership became a central policy for Thatcher's government, and so, alongside council house sales and the 1981 monetarist tax budget, BT's privatization became one of Thatcherism's "three pillars."[22] BT's privatization typifies Thatcherism, which included both pragmatic and ideological elements.[23] At first more reactive and antisocialist, and not self-consciously neoliberal, Thatcherism became, after BT's sale, a more prescriptive and market liberal project that oversaw the retreat of the state from public ownership, one of the most significant changes to the structure of the British economy in history.[24] In short, Thatcherism would not have existed without BT's privatization.

BT's privatization influenced politics abroad, not just domestic politics. Thatcher's privatization program was, "without question," the most significant privatization program in history.[25] BT's sale persuaded other countries to sell their state-owned enterprises, and privatization established new ways of thinking about the demarcation between the public and private sectors.[26] Privatization was a popular policy export, as British politicians and think tanks toured the world, especially Eastern Europe, promoting and implementing privatization in the late 1980s and 1990s.[27] Histories of privatization have highlighted its role as a core neoliberal policy.[28] BT's privatization was thus a key moment in the "market turn" that started in the 1970s.[29] This market turn is thus an important topic for this book, as understanding it helps better place BT's privatization within changes in British and global political economy. The market turn is generally taken as a shift away from social democracy and toward neoliberalism, but both these terms suffer overly generalized and contested definitions, which can reduce their analytical value, and so they require elaboration.[30]

There are two broad approaches to neoliberalism. The first focusses on neoliberalism as a transatlantic intellectual movement and doctrine that became prominent after World War II, mainly associated with the Mont Pelerin Society, founded in 1947, whose thinkers promoted the free movement of capital and sought to construct competitive markets as a superior mode of economic governance, compared with the nation-state.[31] The second approach explores "actually existing" neoliberalism, investigating privatization, deregulation, and other local and various neoliberal practices from the 1970s onward, a trend called both "neoliberalization" and "market liberalism."[32] Critiques of neoliberal scholarship argue that, while there are significant connections between these practices and the neoliberal intellectual movement, these connections are not linear and sometimes exaggerated.[33] BT's sale also complicates this dualistic, linear understanding of neoliberalism. The sale radicalized Thatcherism, demonstrating that neoliberalism was never a purely top-down project that moved linearly from ideation to ideology to policy, but that policies such as privatization helped create neoliberal politics.[34] In other words, neoliberalism is "embedded" in socio-material conditions—institutions, industries, and infrastructures—that were important to its success.[35]

While neoliberalism is complex, social democracy is potentially more problematic as an analytical term for this history. Social democracy originated in early twentieth-century Europe as a democratic alternative to revolutionary socialism. Social democrats sought to use the democratic state to promote social solidarity and to institutionalize policies based on progressive taxation, guaranteeing economic security by funding social insurance, education, healthcare, housing, and infrastructure.[36] But it would be wrong to assume that postwar, pre-neoliberal UK political economy was social democratic. Britain's twentieth century, especially after World War II, was defined by the rise and fall of a nationalist political economy.[37] In this period, Labour was more the party of economic nationalism than of social democracy, while the Conservative Party mixed competing interests in empire, nation, free enterprise, and free trade. From the 1940s to the 1970s, this nationalist political economy most defined the ownership and governance of Britain's economy, and from the 1980s, the Thatcher governments swapped economic nationalism for internationalism, with privatization as one of their chief methods.[38]

This means that, for most of the period that this book covers, from the 1950s through to the 1980s, British telecommunications was not a social democratic or welfarist public service, but an instrument of nationalist political economy. This also helps understand privatization as more than merely an act and agent of neoliberal political economy. Multiple politics pervaded privatization. Before it was called "privatization," selling state-owned enterprises was called "denationalization." This term better draws attention to the fact that selling state-owned enterprises also represented the decline of nationalist political economies. With Thatcher's rise to power came a contradictory mixture of waning nationalism and burgeoning neoliberalism. Alongside this, international finance asserted itself more in the British economy by purchasing state-owned enterprises, as happened with BT, approximately one-third of which sold to foreign investors. Privatization was more a move away from a nationalized political economy than it was a move away from social democracy.

Privatization is thus best understood "from an eclectic perspective," wherein the whole is more than the sum of its parts.[39] Studying its history solely from the perspective of pragmatism, Thatcherism, neoliberalism, or nationalism misses the analytical richness that it offers as a case study of how all these political economies intersected. Infrastructure privatization and deregulation have a complex history, which promoters and policy analysts tend to miss in favor of promoting simplistic narratives.[40] In particular, the role of technology is either overlooked or treated superficially. From the late

1980s and 1990s, policy scholars studied telecom privatization and deregulation in Britain, Europe, and the United States.[41] Their accounts saw deregulation and privatization either as ideological or as enabled by exogenous technological change—or as a combination of the two. They also often cited national institutional factors to explain the routes that different nations took toward telecom deregulation and privatization. This mirrors a key move in histories of neoliberalism, which have recognized how national institutional conditions shape ideology, showing how "policy creates politics."[42] BT's sale is a quintessential example—a policy that radicalized Thatcherism. But these histories and policy studies have too restrictive a view of the diverse institutions that shaped privatization and neoliberalism. These infrastructures are themselves socio-material institutions, and technology was not an exogenous pressure on national infrastructure, but rather something that engineers developed and innovated through these infrastructures.[43] Infrastructures "act like laws" in setting the limits of possibility, meaning that scholars must pay more attention to how engineers, technocrats, and their material contexts shaped the political economy of privatization.[44] This means understanding the technological landscape before privatization, requiring a longer history of what digitalization looked like under national ownership.

DIGITALIZATION AND NETWORKING: BUSINESS, POLITICS, AND CONVERGENCE

The digitalization of communications infrastructure was not a process that developed in a vacuum but instead built on telecommunications networks that had existed for a century. Complex public-private relationships had characterized these network businesses long before BT's privatization. In North America, there was a long and heated debate across the nineteenth and early twentieth centuries over whether telecommunication was an essential public service.[45] From the 1840s, a state-oriented political economy encouraged competition between private corporations, but in the 1880s, a municipalist political economy that idealized public utility meant that many cities consolidated these corporations into independent public utilities. This was the era of the independent public phone company, showing a viable alternative political economy of telecommunications. At the turn of the century, however, a new progressive political economy, which believed that national monopoly would reduce waste and duplication, undermined

the independent telecom movement. Combined with the political adroit-
ness of AT&T's leaders, who presented their private telecom monopoly as
a "universal service" that brought telecommunications to the nation, the
independent movement ended. The private Bell System monopoly became
the dominant North American telecom provider.

In Europe, there were many different paths to national telecom systems.
Some countries, such as Germany, were state-led from the start, while others,
such as the Scandinavian countries, had a more open approach that combined private capital and public investment. The United Kingdom, Spain,
and Italy took a middle ground, where private firms financed and provided
telecom infrastructures, but these were tightly regulated and often nationalized by the end of the nineteenth century.[46] There was no single path to
monopoly, and instead, each nation, when confronted with new technologies, determined the boundary between public and private in different ways.[47]
Regardless of the path taken, however, by the early twentieth century, the
dominant organizational model in Europe was the PTT system, where telegraphy and telephony were offered alongside postal services by one business.
In the UK, this was the General Post Office (GPO), a Civil Service department.

War and wireless shaped the business and politics of national networks in
the first half of the twentieth century. Radio, known as wireless telegraphy,
challenged state monopolies by evading the regulatory frameworks governing traditional wired communications.[48] European states moved to regulate
wireless quickly, particularly as radio became the first electronic mass media.
Radio unsettled the political economy of communications, with different
responses in different nations. The UK created the BBC, while in Germany,
international wireless communications became an opportunity to challenge
British imperial hegemony over global communications.[49] World War I also
influenced the political economy of communications. Military development
laid the groundwork for the interwar European electronics industry, accelerating wireless technology.[50] Furthermore, the US entry into World War I created exchanges between North American and European political economies
of telecommunications.[51] Bell engineers helped build the US Signal Corps'
European wartime networks, which AT&T used to reaffirm its commitment
to public service, turning sentiment against the brief nationalization of US
telecom in 1918. Contact with Bell engineers also empowered French critics
of state-owned networks and led to a French-American alliance that gave Bell
a foothold in the European telecom industry. Similarly, World War II also

shaped the political economy of European telecommunications, as the cooperative spirit of "technocratic internationalism," born in nineteenth-century
Europe, not only survived but thrived during the war, cementing infrastructural Europeanism after the war's end.[52]

Just as World War I matured wireless technology, so World War II stimu-
lated digitalization. Ballistics, code-breaking, and nuclear physics led to computers, the first class of digital machines.[53] The growth of a global computing
industry took place in several waves, starting from its origins immediately
after World War II, followed by its standardization in the late 1960s and
1970s.[54] Personal computing appeared in the 1980s, followed by the internet
and web in the 1990s. This last wave was the era of "digital convergence,"
when computing and communications combined to form a global digital
network. Digital convergence, however, as this book and others show, has a
much longer history. From the 1960s, North American and European users
could remotely and simultaneously access computers for various resources,
such as processing power, using a service called time-sharing, which was
a major sector of the computer services industry.[55] Alongside important
services like time-sharing, various themes characterize this history of digitalization: miniaturization, standardization, systematization, and competition
among different technical and national visions of computing.[56] As this last
theme suggests, these were not purely technical issues and intersected with
political economy in different ways around the world.

In the US, computer firms were emblematic of the "corporate commonwealth," a new postwar political economy.[57] Although many of these firms
had roots in the prewar office appliance and electronics industries, after
World War II, they both built and thrived on a political economy characterized by Cold War government spending, US engagement with the wider
world—not least via state support of US exports—and corporate "welfare
capitalist" packages designed to combat unionization. The iconic example
of this is IBM, which profoundly influenced both international digitalization and US political economy. IBM dominated the US computer market
to such an extent that it was frequently subject to antitrust action, particularly around its practice of "bundling" hardware, software, and computer
services, such as training, together. From 1969 to 1982, the US Department
of Justice ran the longest antitrust case in US history against IBM over this
practice. IBM "unbundled" its software products and computer services pre-
emptively, in 1970. This allowed customers to buy software and computer

services from independent companies, separate from leasing or purchasing IBM hardware, and gave wings to a fledgling software products industry.[58] The US courts eventually dismissed the Department of Justice's case as without merit in 1982, and the case soon became a symbol of the government's wastefulness and unwarranted intrusion into the private sector, reinforcing the rise of a deregulatory political economy under Ronald Reagan.[59]

Outside the US, digitalization and political economy blended in different ways. Most prominent in Europe was a "national champions" policy, particularly in the UK and France, which sought to create national computing industries and domestic champions to compete with IBM.[60] In West Germany, on the other hand, IBM had a powerful influence on ideas about productivity, labor relations, and the role of corporations in a capitalist, growth-oriented economy.[61] The national champions policies in the UK and France were not successful. France's national champion, CII, never became successful enough to sustain a domestic electronics manufacturing industry and was bought by Honeywell-Bull in 1975. The growing standardization of international computing and fraught relations with government dampened a defiant British exceptionalist approach in its national computing industry and its champion, ICL.[62] In the UK, these efforts to centralize and control computing also resulted from and failed because of gendered, classist labor hierarchies within both government and the computing industry. Managers favored large, centralized mainframes that disempowered technical workers, often women, just as the international computing industry was moving away from mainframes and toward minicomputers.[63] This relationship between British government and computing went far beyond interventionism. The British "government machine" had long been concerned with information technologies as ways of discreetly centralizing the bureaucratic state.[64] Technocratic experts so thoroughly wove computing into state bureaucracy that the structure of government itself followed computing trends, from centralized in the mainframe era to outsourced in the era of networked personal computing.

This history is not purely about computing, however. It is also about digital networks. From the 1960s, data communications networks began to appear, particularly in the US, which eventually formed the foundation for the internet.[65] But while the internet was emerging in the US, other countries were also building alternative digital networks. France created a remarkably successful online digital network, Minitel, using its national telecom infrastructure. Minitel ran from 1980 to 2012 and showed an alternative model for digital infrastructure based on creative and collaborative public-private partnerships.[66] The USSR, on the other hand, failed to build a successful national digital infrastructure. Intense competition among different ministries hamstrung the development of OGAS, the All-State Automated System, compared to the "cooperative capitalist" approach that sustained US internet development.[67] Chile is perhaps the best example of a successful socialist digital infrastructure.[68] Between 1971 and 1973, the socialist government of Salvador Allende built Project Cybersyn to provide real-time control over the Chilean economy by networking industry and government. Cybersyn helped the Allende government overcome a national trucking strike in 1972. Cybersyn, however, was frequently cast as a form of totalitarian control inside and outside Chile, and the Pinochet government shut it down after the 1973 military coup. Japan's influential Fifth Generation Computer Systems project represented another alternative political economy of digitalization.[69] This project, begun in 1982, was the first national large-scale artificial intelligence research and development project free from military influence and corporate profit motives. FGCS was still state-led but, unlike AI research in the US, was oriented to socially responsible innovation rather than Cold War America's militarist-capitalist ambitions.

France, Chile, and Japan show that alternative political economies of digitalization are possible, but this only reinforces that one mainstream digital ideology has emerged from the US. Utopian narratives about a digital future and the power of systems originated from early US digital research after World War II.[70] By the 1980s, this had solidified into an "information age" discourse that emphasized information technologies' centrality to society and the economy.[71] Margaret Thatcher's labelling of 1982 as IT-82 was one such act that popularized the information age. The information age has also arrived packaged with a utopian ideology about the liberating qualities of digital technology. This ideology's rise explains how, from the 1950s to the 1980s, the computer as a symbol has moved from big, centralized, and bureaucratic to small, networked, and personal, facilitating freedom of choice and expression for a market economy.[72] Different labels have emerged for this digital ideology. "Mythinformation" denotes the belief that information technology would lead to more free and democratic societies. The "Californian ideology" describes a US West Coast ideology that saw digital technology as fusing countercultural values of individual freedom with Silicon Valley's penchant

for economic liberalism. "Digital utopianism," associated with groups ranging from the New Communalist movement to *WIRED* magazine, promoted
digital technologies' capacity to beget harmony, egalitarianism, and freedom.[73] There were some exceptions to these mythologies: for example, from
1965 to 1975, the success of time-sharing networks led to extensive calls in
the US for computing as a public utility, idealizing communal, civic values.
These values were embedded in various time-sharing networks, such as the
Minnesota Educational Computing Consortium, a statewide public computing utility, but were eclipsed by the rise of personal computing.[74] In general,
however, histories of digital ideologies highlight the engineering values and
choices that have formed "the defining libertarian mythology of Internet
culture."[75]

These ideologies propagated beyond the US, but it would be a mistake to
see them as sovereign. The changes that took place through digitalization
were also shaped by national policymaking and the internationalization of
telecommunications markets. For example, from the 1970s, digital network
builders promoted an "open ideology" that was skeptical of concentrated
power in closed systems and instead placed faith in the combined power of
market capitalism and technical expertise.[76] These libertarian values fueled
experiments in French networking, and by the 1980s, the principle of openness and open networks became a cornerstone of European telecommunications competition policy as it moved away from the PTT monopoly system.[77]
But while the open ideology was influential, and perhaps even necessary, it
was not sufficient for changes in European telecommunications competition
policy. These changes also emerged as a competitive response to the deregulation of US and UK telecommunications.[78] The divestiture of AT&T in the
US was not just about breaking up a private monopoly, but also about supporting AT&T's expansion into international communications markets and
services.[79] In the UK, pressure from organized business, especially finance, for
international communications services also motivated BT's privatization.[80] In
turn, the competitive threat of BT and AT&T expanding into international
markets added further incentive for European nations to deregulate.[81] It was
thus the intersection of privatization, digitalization, and market internationalization—in other words, policy, technology, and commerce—that made
these transformations.

These changes show that neither digital nor neoliberal ideology alone led to
the present condition of modern communications infrastructure. Ideological

histories of digitalization can be subverted and complicated by finding the "missing narratives" of digital networks, which should move beyond histories of the internet to consider the broader history of digital networking. For example, there are few "telephonic histories of digital technology."[82] These histories should also treat digitalization as an analytical category, particularly during the "early digital," which means taking seriously the history of digitalization beyond the computer and exploring digitalization in different sectors, particularly state institutions.[83] The accomplishments demonstrated by the histories of the Minitel in France, Cybersyn in Chile, and OGAS in the USSR challenge the libertarian ideology of the internet by showing how different actors in different nations enacted digitalization. They show that alternatives were and are possible.

But one final missing narrative requires further investigation, and that is how digitalization has affected major transformations such as neoliberalism.[84] Illuminating that influence is one of the key goals of this book. Britain's telecom infrastructure was a non-internet digital network that was shaped by national ownership and yet was profoundly important to the success of privatization as a key piece in the neoliberal policy package. How did its managers and engineers make and remake their network as they turned from their priorities as a national monopoly to new priorities of privatization and market liberalization? Exploring that question requires a finer conceptualization of the relationship between technology and political economy, to which this introduction now turns.

HISTORY, TECHNOLOGY, POLITICAL ECONOMY

The first way to understand the relationship between technology and political economy is to see that technological and infrastructural change can reflect dominant political economies. This chapter already describes numerous examples. The organization of North American telecom infrastructure followed the shift between state-oriented, municipalist, and progressive political economies.[85] In the postwar US, a strong regulatory political economy led to the unbundling of software and computer services.[86] Project Cybersyn in Chile followed a democratic socialist political economy, while Japan's Fifth Generation Computer System followed a state-led public-good-oriented political economy.[87] Examples extend beyond digital and communications technologies too. The reform of UK postal infrastructure in the nineteenth century followed a market liberal political economy, while similar US reforms followed a civic nation-building republican political economy.[88] In the late nineteenth and early twentieth centuries, electrical network construction in Berlin, Chicago, and London followed local political economies.[89] Berlin's mixed political economy, balancing public interest and private enterprise, led to an advanced private infrastructure that was eventually taken into public ownership. Chicago's ambiguous and weak regulatory regime gave private enterprise much freer reign, while London's devolved, localist political economy created a fragmented electrical network.

But to view infrastructural change as following political economy misses the critical historical insight that infrastructure has co-constructed political economy. Infrastructure inspired some of the earliest theories of political economy. The development of London's water supply influenced J. S. Mill and Nassau W. Senior's theories of ownership and market structure, while US economist Richard T. Ely's concept of natural monopoly was inspired by his understanding of rail and other network infrastructures.[90] In the US, rail influenced more than just public economy theory. Rail infrastructure also led to new trade associations and cooperative industry agreements that bypassed the existing open market patent system, thus becoming the basis for a new corporate political economy that displaced a market economy.[91] Not just in the United States, but also in Sweden, rail became a "paradigmatic system" that influenced political and economic thought about infrastructure ownership and governance.[92] In Britain, the birth of modern infrastructure, in the form of road building in the eighteenth and nineteenth centuries, created an engineering elite, a modern national bureaucracy, a new national political economy, and the political philosophies of J. S. Mill and Adam Smith.[93] Even at the supranational level, this has been this case. For example, there could be no political economy of Europe without the transnational transport and communications infrastructure that made Europe an economic unit in the first place.[94]

This co-construction of technology and political economy has defined the twentieth century. The first modern social insurance systems in Germany and Switzerland, central to the formation of the welfare state, developed in mutual relationship with actuarial science as a "technology of trust" that made social insurance possible.[95] In the US, the New Deal and the construction of a federal highway infrastructure relied on an ideology of technical expertise. Planning tools from the Cold War military-industrial complex, such as computer simulation and satellite reconnaissance, shaped welfarist postwar urban planning.[96] In the UK, 1970s urban infrastructure projects demonstrated the dynamism of late social democracy before contesting and adapting to the market turn.[97] Neoliberalism appears to be a political economy that thrives off a mutually productive relationship with science and technology.[98] For example, computers and information theory influenced neoliberal economists' view of the market as an information processor.[99] The West's transition from coal to oil as its main energy source also saw a shift away from the mass politics associated with the coal industry's labor force and toward a liberal market politics based on the apparently unlimited growth gifted by foreign oil.[100] Similarly, nationalist political economies have often been techno-centric. Technologies such as nuclear power and airplanes have fueled nationalist politics, and techno-nationalism pervades state-oriented political economies of innovation.[101]

These examples show the "technopolitics" of the modern age.[102] Technopolitics views technological and political life as mutually constitutive, describing the ways that technology can both reflect and enact political goals and, furthermore, conceal these goals' material origins. Politics often draws a hard line between human and nonhuman, presenting the material world as subject to human intentions rather than something that can shape human politics.[103] Technopolitics can thus show alternative political economies of technology, but it can also produce an "anti-political economy," meaning that, as technology sets limits to politics, so technopolitics can close the space of political and economic possibilities.[104] This is something that histories of infrastructure privatization miss by not exploring how the technopolitics of infrastructure have shaped nationalization and privatization. This book therefore demonstrates, first, how digitalization shaped and was shaped by national ownership, and second, how this process changed with liberalization and privatization.

Focusing on digital infrastructure rather than a specific digital technology helps achieve these goals. Infrastructure history is a history neither from above nor from below, and as a subject that exists at all social scales, it can challenge the tinted lenses of histories at the micro, meso, or macro scales.[105] It offers an understanding of the messy, technical complexities of national ownership, privatization, and digitalization beyond politics and ideology. This approach is an example of how infrastructure history "inverts" traditional historical narratives. Paying attention to the low-level technical practices that compose infrastructure, such as standardization, gives greater insight into the changes that preceded and enabled apparent technopolitical transformations, such as digitalization and privatization.[106] BT's privatization is one such breakthrough narrative, cast as a landmark moment for neoliberalism and used in the UK to tie market practices to the "information age." In contrast, infrastructural inversion highlights the need to explore the practices that preceded and enabled BT's sale. By drawing attention to the practices that have been central to anticipating and enacting broad social, political, and technological change, infrastructural history can recover the technopolitics that "breakthrough" narratives such as privatization overwrite.

Focusing on these practices means focusing not on a monolithic infrastructure, but on the actors that made and ran that infrastructure. The history of technology has deep roots as a history of engineers and "system builders."[107] Likewise, the study of management and managers has been foundational for business history.[108] But what these histories often show is that managerial and technological changes are interdependent.[109] Clerical technologies, such as typewriters and computers, shaped new administrative theories and practices in US business and British government.[110] Similarly, the success of system-builders in various infrastructures was contingent not just on their technical vision but also on their managerial prowess and ability to mobilize political and financial resources.[111] The mutual dependence of managerial and technical expertise has had wider implications for modern history. The systematization of management in the late twentieth century has been labeled the "new spirit of capitalism," while technoscientific expertise has supported the material and administrative integration of Europe and the "locationless logic" of colonial administration.[112] These technocrats, especially those in the public sector, are thus best seen as "heterogeneous engineers," combining innovation, management, and government into a greater practice, a practice that requires the engineering not just of artifacts but also of people, institutions, and ideas.[113]

British telecom infrastructure offers an ideal case study of these people and practices that made and remade digitalization, nationalization, and privatization. Britain's market turn was an international moment, both in the ways that global processes contributed to it and in the ways that it traveled around the world.[114] BT's privatization exemplifies this, a proof-of-principle for a privatization movement that went global, creating opportunities for international finance and challenging Europe to deregulate or be left behind.[115] Britain's telecom infrastructure was local, a public infrastructure run for, by, and within Britain, and yet also networked, both materially and socially, to a wider world of telecom infrastructure and ideas about how to run that infrastructure. Its history can thus give new insight into the late twentieth century's political transformations. Furthermore, Britain provides an alternative history to those histories of digitalization that mainly focus on the US and the influences of libertarianism and the Cold War's "closed world" on the histories of early data and internet infrastructure.[116] Britain's telecom infrastructure ran on the European postal, telegraph, and telephone (PTT) model, and so this history provides an opportunity to expand on the European model of integrating new data infrastructures into the existing infrastructural pattern of national ownership.[117] Just as histories of digitalization in Chile, France, and Japan offer insights into alternative political economies of digitalization, so can a history of the digitalization of Britain's telecom infrastructure. But in this case, this history also shows how digitalization intersected with British Telecom's privatization, one of the most politically significant moments in the global twentieth-century history of infrastructure.

BRITAIN'S TELECOM INFRASTRUCTURE

The existing historiography of Britain's telecom infrastructure neglects technology for political and economic questions about monopoly and market, public and private. In this narrative, the British government, especially the Treasury, stifled infrastructure through spending restrictions, trying to have its cake and eat it by operating a public utility without providing the necessary costs of upkeep and investment. This narrative of conflict between the "dead hand" of Treasury "restrictionists" and the Post Office "expansionists" echoes an old historiography of modern British state spending, which focused on how much money the state spent, rather than on what the state spent its money.[118] This history is dangerous for two reasons. First, it reinforces an outdated picture of the "dead hand" of the Treasury, a stereotype that has less basis in reality and more in twentieth-century Britain's politics of "declinism," and misses that the Treasury was far more dynamic than remembered, especially when it came to information technology.[119] Second, it flattens the history of nationalization and privatization into a calculus of capital, where neoliberal fiscal restraint waited in the wings to replace an expansive but investment-starved welfare state. In the case of the Post Office, a key unanswered question is not whether the Post Office struggled with the Treasury's fiscal restraint, which is already clear, but how the Post Office directed its resources, in both money and personnel, to some technologies and not others.[120]

When this story starts, around 1950, the British telecommunications infrastructure was a state-owned monopoly run by the British Post Office. The Post Office had run telegraphy since the 1869 Telegraph Act, written in response to a telegraphic price-fixing cartel. The 1869 act gave the Post Office an indefinite monopoly over electrically carried communications, but did not expand into telephony because the Treasury refused to authorize spending and, apparently, because Thomas Edison failed to impress the Post Office's engineer-in-chief, William Preece, with a demonstration of the telephone.[121] By 1912, public and political dissatisfaction with the National Telephone Company (NTC), a private monopoly, led to the Post Office purchasing the NTC's assets, completing the British state's takeover of telecommunications.[122] The Post Office inherited a system suffering from chronic underinvestment, but Treasury spending controls slowed investment. Political pressure meant that, in 1932, the Post Office and Treasury negotiated a new financial arrangement whereby the Post Office paid the Treasury a fixed annual sum of £10.75 million (about £73 million today) and could reinvest its excess profits.[123] This arrangement lasted until World War II, when the Treasury resumed direct control of the Post Office's finances and refused to revert to the former arrangement after the war ended.

By the beginning of the 1960s, the waiting list for a telephone connection had ballooned.[124] Harold Macmillan's Conservative government thus passed the 1961 Post Office Act to reinstitute the prewar fixed annual payment arrangement, which had been back on trial since 1956. This arrangement had little effect, as successive Treasury capital restrictions in 1962 and 1963 limited Post Office borrowing. The Labour prime minister Harold Wilson's 1966 July measures, created to avoid sterling devaluation, further cut telecom investment.[125] Through the 1960s, calls therefore came for more freedom and a more commercial attitude in telecommunications. Perhaps loudest was Tony Benn, appointed as a young, modernizing postmaster general in October 1964 after Labour's election victory.[126] Benn's top priority was turning the Post Office from a Civil Service department into a public corporation, known as the "break," and separating telecoms from post, called the "split." In August 1966, Edward Short, Benn's replacement as postmaster general after Benn was appointed minister of technology, announced the "break," but the "split," believed impractical, never happened. On October 1, 1969, the Post Office Corporation was formally created and, as this book will show, this shift to a more corporate structure, attentive to commercial trends, particularly in business and finance, would shape the Post Office's plans for public digital infrastructure.

The 1970s brought tension between the Post Office's commitment to building this digital infrastructure and the turbulent political and economic environment. Spiraling inflation in the early 1970s meant that the Post Office had to balance high wages, voluntary price restraints set by the Confederation of British Industry, and raising postal and telephone charges to meet the government's financial objectives.[127] Alongside this, the Treasury used the Post Office as an "instrument of macro-economic management," cutting spending by £150 million, followed by further reductions in 1975–1976 as telephone demand dropped because of the 1973–1975 recession, and another cut followed the Winter of Discontent in 1979.[128] Despite this, telecom was one of the nationalized industries' better performers across the 1970s, but increasing waiting times and the burden of subsidizing the postal business meant calls for the split grew, including from William Ryland, the Post Office chairman, and the Post Office board. In 1975, Harold Wilson's new Labour government thus initiated a review, the Carter Committee, into the Post Office's structure. This committee recommended in 1977 that the government separate post and telecom.[129] The Labour government, however, rejected this recommendation, instead using the Post Office for an experiment in industrial democracy that added trade union and external industry representatives to the Post Office board.[130] This particularly aggrieved the Post Office's new chairman, William Barlow, appointed in 1977 under the assumption that he would oversee the split and take charge of telecom.

Shortly after the Conservatives won the 1979 election, Barlow thus attacked the new government on two fronts, threatening his resignation and openly calling for privatization.[131] Keith Joseph, the secretary of state for trade and industry, quickly announced an end to the industrial democracy experiment, the split of post and telecoms, the creation of BT, and a review of the telecom monopoly, setting the stage for liberalization.[132] The government introduced the Telecom Bill in 1980 to turn BT into a separate corporation, leaving the monopoly mostly intact but granting the government powers to license new operators and launching competition in customer equipment and private exchange supply and maintenance.[133] The bill gained royal assent in July 1981, and BT formally opened for business on October 1, 1981. Meanwhile, the government also commissioned a report into telecom liberalization. This report, published in April 1981, recommended that BT should lease circuits to competitors to resell voice telephony; that the government liberalize value-added network services, where third parties could lease lines for non-voice services like electronic funds transfer; and that the government should license new competing telecom systems with separate physical infrastructure.[134] Strong BT opposition and Home Office security concerns meant that the government chose a compromise in which it accepted proposals for a single alternative network, which eventually formed in June 1981 as Mercury Communications, creating a duopoly system that lasted until 1991.

Telecom underinvestment and performance was also, alongside the government's monetarist budget, a significant factor behind BT's privatization.[135] Telecom's self-financed investment had risen from the mid-1970s, but the Post Office's projected costs for a public digital infrastructure meant that it would need external financing. These costs alarmed the Treasury, given both the new Thatcher government's objective to lower the budget deficit and the external financing limits placed on nationalized industries since the 1976 IMF bailout. The Post Office added further pressure by threatening to add a surcharge to telephone bills to stay within the external financing limits. This pressure only grew as Post Office Telecom became BT, raised prices in November 1981, and exceeded its 1981–1982 external financing limits by 30 percent. This incensed Thatcher, and consumers, who had some of the highest telephone charges in Europe. The government thus began to explore new ways to finance telecom, including privatization, which gained support from BT's chairman, George Jefferson, who had tired of Treasury restrictions. By summer 1982, this had turned from studying a minority sale into a majority flotation that would free BT from the budget deficit and Treasury controls. In July 1982, seventy years after the Post Office took over the British telephone infrastructure, Patrick Jenkin, the secretary of state for industry, announced that the government would privatize BT so that it could raise private capital free from government interference.[136] Two years later, in 1984, BT became a privately owned corporation with a new agenda for its nascent digital infrastructure.

This history of Britain's telecom infrastructure gives some insight into the changes that happened, from nationalization to corporatization to

privatization. It is, however, only a partial history, because technology
remains invisible. As this chapter has argued, however, the history of technology, particularly digitalization, in Britain's telecom infrastructure has
played a crucial role in the history of both the telecom business and the
wider political economy. The birth of plans for public digital infrastructure
in the late 1960s suggests that, while the British state may have been cash-
strapped, it was not bereft of new ideas and projects. These plans also motivated the Post Office board's support for telecom independence in 1976 and
its search for external financing in the 1980s. These moments show how the
traditional narrative, which reduces the history of telecom infrastructure to
a story of government spending, flattens digitalization and national ownership's histories and horizons. BT's sale was not just about public and private,
nor just about monopolies and markets. BT's sale also overwrote a history
of nationalized digital infrastructure, a history that can enliven the present.

OVERVIEW

This book draws on sources from British Telecom Archives, the British Postal
Museum and Archive, the UK's National Archives, the Bank of England
Archives, the Smithsonian Institution, and George Washington University's
Special Collections Center. The book has three parts. Part I, "Plans," investigates how the British Post Office created and sustained its plans for a new
national digital infrastructure. Part II, "Projects," explores how engineers
and managers built those plans into the network. Part III, "Places," explores
how Britain's telecom infrastructure connected local and global through
three different sites.

Part I's first chapter begins with the birth of plans for a national digital
infrastructure in the late 1950s and 1960s. The chapter explores the "information age" discourse that forged these plans and the influence of cybernetics
and information theory on the Post Office. In doing so, it links work on the
influence of cybernetics and information theory with scholarship that highlights plans, visions, and expectations as rich evidence for the dynamism and
persistence of specific ideas.[137] Chapter 2 continues this theme by tracing the
history of the Post Office's Long Range Planning Department, the telecom
infrastructure's visioneering and forecasting unit, from its founding in the
late 1960s to its role in BT's liberalization and privatization in the early 1980s.
Founded during a broader trend for futurology and futures studies in business

and government across the Western world, this department sustained the Post Office's plans to "invent the future" for digital infrastructure.[138] Chapter 2 follows these practices through liberalization, as the department went from inventing digital futures to forecasting market futures, using corporate modeling and simulations, which informed BT management's plans to use predictive computing to surveille and simulate customers' total information needs. In doing so, this chapter shows how "prediction technologies" have been crucial to marketization.

Part II turns to how, from the 1960s, the Post Office and BT embedded these plans into infrastructure. This part looks at two basic features of a telecom network, switching and transmission, and how digitalization affected these components. Part II first looks, in chapter 3, at switching's automation and computerization, and how that influenced the Post Office's monopsony over its equipment suppliers and telephone exchange labor pools. Chapter 4 looks at the myth that Margaret Thatcher killed BT's plans for a national fiber-optic network. The discussion sets this myth in the context of the Post Office's vision of an "integrated" digital network that carried both telecommunications and video services, including television. The Post Office searched for both transmission technologies and standards that would support this vision. Chapter 4 shows how, in this search, Post Office and BT engineers were developing more for their expectation that they would inherit TV broadcast responsibilities, rather than for the reality of their actual national responsibilities. Thatcher's "killing" of the national fiber-optic network was more about blocking BT's expansion into the private broadcasting market, than it was about defunding telecommunications.

Part III looks at British telecom infrastructure's local and global sites through chapters on its rural Suffolk telecom laboratory, transatlantic infrastructure, and financial users in the City of London. Chapters 5, 6, and 7 build on approaches that emphasize privatization and infrastructure as both local and global and that show how material and ideological networks are made global through local processes.[139] By studying these sites, this book shows how the digital and market turns of Britain's telecom infrastructure resonated globally. Chapter 5 investigates the Post Office's new research laboratory in Martlesham Heath, East Suffolk, from the late 1960s. The chapter shows the transformation of a modern state-owned research site, influenced by trips to US corporate research campuses, into a postmodern privatized information technology park, alongside the construction of an anachronistic "new village" around the research site. The relationship between local and global at Martlesham Heath reveals how privatization and digitalization invoked both spatiality and temporality. Chapter 6 looks at how the Post Office and BT collaborated with AT&T to lay submarine communication cables and how this effort was threatened by communications satellites. It situates this infrastructure-building within the Post Office and BT's ambitions to expand into international markets, showing how these ambitions prefigured BT's liberalization and privatization under Thatcher. Chapter 7 explores the role of financial users in the City of London in BT's privatization. It shows how the Bank of England and City users formed a lobbying group in the late 1960s to secure priority for the financial sector in Britain's telecom infrastructure, and how the Post Office and BT became initially reluctant, but later willing, players in the ambition to develop London as the world's capital of finance. Altogether, this section shows how rural Suffolk, the City of London, the Atlantic Ocean, and even outer space became contested sites for the market turn and the information age.

The book concludes by returning to its overarching aims of recovering alternative political economies of digitalization that existed under national ownership and showing how that changed with privatization and liberalization. This book shows new ways to understand the digital and market turns, but more than that, it offers lessons for new plans to build public digital infrastructures.

I PLANS

1 THE ORIGINS OF THE DIGITAL VISION

In October 1967, James Merriman, the Post Office's senior director for engineering, outlined a vision for a new digital infrastructure for Britain. In a speech titled "Men, Circuits and Systems in Telecommunications," his inaugural address as chairman of the Electronics Division of Britain's Institute of Electrical Engineers, Merriman described a "general-purpose digital network" that would "freely handle all forms of communication," so giving Britain a public digital infrastructure that would provide voice, data, and video services.[1] In this vision, Merriman blurred the line between the machines that composed this infrastructure and the people who ran and used it. He discussed the interplay between "circuits and men," arguing that telecommunications was "not only a dialogue between sender and receiver, it is a dialogue between user and provider. It is not only circuitry; it is man management."[2] Digitalization would produce a "self-healing, self-governing" network that could administer this management. For the Post Office, digitalization was not only about services but also about management, and would unify information and administration. In this vein, Merriman concluded his speech by arguing that "information and control" were "fundamental to any telecommunication system."[3] Merriman's speech announced the Post Office's plans to build a comprehensive, nationally owned digital infrastructure for Britain and showed an expansive vision of what digitalization meant. This chapter traces this vision's origins from an analogue failure and its influences, including cybernetics, managerialism, the British "government machine," and the threat of independent data networks.

Failures and visions drove this intersection of technology and politics. The history of technology has long shown that focusing on success alone makes for Whiggish stories of technological progress and that understanding failure reveals much about the priorities and expectations of those developing and using those technologies.[4] These visions and expectations, marshaling resources and guiding development, are as central to the history of technology as material artifacts themselves.[5] Failures can be fertile soil for both old and new visions, nurturing an old vision's quixotic status or giving room for a new vision to grow. This does not mean that failure is a prerequisite for success, but rather that the perception of failure and its circumstances shapes the ground from which subsequent visions and expectations can grow. This dynamic between failures and visions helps understand not just the technological history of Merriman's vision, but its political history too. Merriman's digital vision also grew from the belief that British government and the Post Office were not technocratic or managerial enough, and it was these failings, alongside the expectations set by cybernetics and new data technologies, that would fuel the Post Office's vision of a universal digital network.

ANALOGUE FAILURE

Through the 1950s and early 1960s, one of the Post Office's main projects was Highgate Wood, an analogue telephone exchange that the Post Office and British government wanted to be the world's first all-electronic operational telephone exchange. The Highgate Wood project officially began in 1956, when the Post Office formed the Joint Electronic Research Committee with its suppliers, Siemens, ATE, Ericsson, GEC, and STC, to coordinate development with industry.[6] Its origins, however, lie with the Post Office's need, following World War II, to upgrade its old, large, and unreliable network of electromechanical Strowger exchanges. Strowger exchanges automatically routed telephone calls using mechanical selector arms that would relay phone calls by connecting to a specific electrical contact in a contact bank. A Post Office engineering delegation thus visited the United States to report on a new version of the Crossbar exchange, an electromechanical alternative to Strowger.[7] Strowger, first patented in the US by Almon Strowger in 1889, had entered service in Britain in 1912 but was no longer up to standard.[8] The delegation reported that Crossbar cost more, used more mechanical relays, and needed more room and power, but concluded that it provided superior service, was more adaptable, and could flexibly serve many locations, from small rural exchanges to metropolitan and trunk exchanges.[9]

These mixed conclusions perhaps suggest the reported conflict between the two lead engineers on the trip, Tommy Flowers and Donovan Barron. Flowers supported moving straight to electronics, while Barron, later engineer-in-chief of the Post Office from 1965 to 1967, advocated Crossbar as an intermediate step.[10] Flowers won, persuading the controller of research, Gordon Radley, that all research and development should focus on an all-electronic exchange. Radley's support, and Flowers's favor for electronics, arose from Flowers's prewar switching research and his wartime creation of Colossus, a digital codebreaking machine developed for use at Bletchley Park, the Government Code and Cypher School. Before the war, Flowers developed an experimental electronic switching device, which the Post Office used minimally from 1939.[11] The war, however, interrupted, and Radley approached Flowers in 1941 to undertake work for Bletchley Park, at which point Radley and Flowers became the first Post Office engineers initiated into British codebreaking at Bletchley Park.[12] Flowers first worked on a special-purpose electromechanical device for Alan Turing's codebreaking team, but after Bletchley Park scrapped that project, Flowers proposed Colossus, an electronic machine to process German messages. Upon completion, Colossus was an immediate success, doubling the codebreakers' output.

Colossus has been touted as the first electronic computer, but it was more influential on Flowers's pursuit of an all-electronic exchange than it was on the first computers. Colossus is better thought of as a digital signals processor and an "exemplary artifact" of the early digital, a period in which diverse experimental digital devices, including the first digital computers, were built in a range of contexts.[13] In Colossus's case, it had a closer relationship to Flowers's work on electronic switching than it did on computing. Flowers's work on Colossus supported his work on Highgate Wood, even though the former was a digital codebreaking machine and the latter an analogue telephone exchange. When Flowers first proposed Colossus, Bletchley Park codebreakers thought it would be too unreliable and that the war would be over before it was ready. Radley, however, allowed Flowers to pursue the project independently at Dollis Hill, the Post Office's research station.[14] This backing continued after the war. In justifying Highgate Wood to Postmaster General Ernest Marples, Radley invoked Dollis Hill's wartime work for Bletchley Park and capitalized on Bletchley Park's computing legacies by telling Marples that "the various projects undertaken by Dollis Hill during the war had led to the idea that electronic techniques could be applied to telephone switching. Somewhat similar techniques have been worked in the development of computers."[15] According to Roy Harris, a member of Flowers's postwar switching team and later one of the chief architects of the Post Office's digital network, Radley always sided with Flowers in electronic exchange development.[16] Radley's support shows how Colossus's later significance lay more in communications than in computing. This support also shows the lasting influence of Britain's "warfare state." Rather than disappearing with the end of World War II, Britain's wartime interests continued to shape state technology projects well into the 1950s and 1960s, with the British welfare state emerging at the end of the 1960s.[17] While Highgate Wood was not a defense project, the Post Office's, and specifically Flowers's, role in wartime codebreaking helped secure political support for Highgate Wood, thus showing the impression that the warfare state left on even the most civilian of projects.

Flowers, however, remained bitter about Colossus's secrecy. Flowers watched on as his Bletchley Park colleagues, the mathematicians Alan Turing and Max Newman, built from the principles of electronic code-breaking to develop computers, while he had "no power or opportunity to use the knowledge effectively. With no administrative or executive powers, I had to convince others, and they would not be convinced. I was one-eyed in the kingdom of the blind."[18] Flowers attributed delays on Highgate Wood to his lack of "prestige, which knowledge of Colossus would amply have provided."[19] Flowers was perhaps disingenuous here, as he had ample support from Radley, the controller of research, and many more factors contributed to Highgate Wood's delays. But Flowers's view that he lacked prestige compared to Turing and Newman had consequences for his relationship with Turing. After the war, Turing worked on a computer project, the Automatic Computing Engine (ACE), for the National Physical Laboratory (NPL), and commissioned Flowers to undertake development work for ACE. ACE would be a stored-program control general-purpose computer used for laboratory calculations, but suffered numerous problems in its development, including Turing's departure from the NPL. Flowers's sluggish attention had delayed ACE, which he had told Turing would be ready by August or September 1946. Flowers neglected ACE and instead focused on updating the oversubscribed postwar telephone network.[20] In effect, Flowers neglected Turing's work on electronic computing for his own work on electronic communications.

From 1947, a small team under Flowers began developing Highgate Wood's basic principles, pulse-amplitude modulation (PAM), which encoded telephone signals as pulsed samples, and time-division multiplexing (TDM), which transmitted signals by rapidly cycling through pulsed samples. Combining these electronic techniques meant that a single telephone circuit could carry multiple signals and that an electronic exchange could route these signals without mechanically moving the circuit.[21] Highgate Wood's development, however, was slow. The Post Office solicited assistance from its suppliers, but the suppliers' concerns about patent exploitation meant that in 1952 they turned down a proposal that had been the outcome of lengthy discussions with the National Research Development Corporation, the Board of Trade, the Ministry of Supply, and the Treasury.[22] Protracted discussions continued until 1956, when Gordon Radley, then director general of the Post Office, told the suppliers that, unless they agreed, the Post Office would choose one of them as its main development and manufacturing partner. The suppliers quickly fell into line, and in May 1956, the Post Office and its suppliers signed the Joint Electronic Research Agreement, which laid out the terms for pooling research, staff, and development tasks. These lengthy negotiations, rather than Flowers's lack of prestige, were likely a greater hindrance to Highgate Wood's development. Flowers and his team had worked mostly undisturbed on electronic switching, and it was only during the 1950s that these delays, caused by manufacturer negotiations, began. Such tensions between collaborative research and competitive procurement were not unique to this period and would continue to shape telecom development into the 1970s and 1980s.

Rather than Flowers's personal prestige, it was national prestige that drove the development of Highgate Wood. In 1957, Ernest Marples, the postmaster general, became concerned about Highgate Wood's delays, which Radley defended as a matter of national interest.[23] Since Flowers and Barron's 1947 visit, Bell Labs had begun developing an all-electronic exchange. In a letter to Marples, Radley pointed to the export prospects for British manufacturing that Highgate Wood offered, reminding Marples of the "resurgence of national competition and nationalistic considerations [which] have tended to restrict the market. . . . The real chance for British manufacturers is to be able to offer electronic equipment that is a stage ahead."[24] National prestige also defined internal concerns about Highgate Wood's troubled development. Two months after Radley's letter to Marples, engineers adjusted

Highgate Wood's power provision estimates. The experimental exchange needed more power, delivered more reliably and stably, raising costs. Lionel Harris, the engineer-in-chief and father of Roy Harris, one of Flowers's team members, raised concerns, and so Donovan Barron, by this point the assistant engineer-in-chief, suggested that they postpone Highgate Wood until a small, economically competitive, practical exchange could be built. Lionel Harris dismissed this idea and "stressed the prestige value of the installation of a complete working system in view of similar projects now being developed abroad."[25] The motivation to beat out foreign competitors such as Bell Labs was clear and urgent.

Highgate Wood's prestige value also fueled additional construction work needed for its growing power requirements. The existing exchange building in north London had to undergo structural alterations, needing an extra room to accommodate more batteries. In turn, this meant that Highgate Wood needed more ventilation through a fan room of at least 1,600 square feet. The existing exchange building had no space for such a room, and so the Post Office requested that the Ministry of Works build an extra floor on the exchange building. In order to justify this, the Post Office explained, "It is a matter of national prestige for us to be the first to introduce such an exchange. . . . It will be of inestimable value to our export trade to be able to be first on the scene with this new development."[26] Despite the engineer-in-chief's earlier acknowledgement that Highgate Wood was an uneconomic experimental exchange, the Post Office still invoked a relationship between research prestige and export prospects.

"Prestige" was a flexible rhetorical device in Highgate Wood's continuing development. This reflects the role that national prestige played in mobilizing British state actors and institutions behind scientific and technological projects in the 1950s. One such example is Jodrell Bank's Mark 1 Telescope, the world's largest steerable radio-telescope dish at the time of its construction.[27] At Jodrell Bank, "prestige" meant different things to different groups. For the Royal Astronomical Society, it meant the standing of British science, while for the Department of Scientific and Industrial Research, it meant a display of British prominence, and for individual astronomers, it was useful as a funding invocation. For Highgate Wood, prestige displayed similar flexibility in coordinating different groups with different goals. For Flowers, prestige meant recognition for his work on Colossus. For the Engineering

Department, prestige meant international recognition. For the postmaster-general and other government departments, prestige meant raising export prospects. Moreover, the ambition that Highgate Wood would be a prestigious world-first is emblematic of contemporaneous British techno-nationalism, which prized scientific and technological world-firsts, such as Comet, the first civilian jet airliner; Bluebird, the speed record-setting land and water vehicles; and Calder Hall, the first commercial nuclear power station.[28]

Highgate Wood's problems, however, delayed the project, which never became a world-first. The exchange entered service in 1962, four years late, two years after Bell Labs' all-electronic exchange, and was an abject failure. Highgate Wood's components failed at a low rate, but the system had several terminal faults.[29] Its connection stores did not always clear pulses after calls, so channels continued to look busy, meaning it would carry less and less traf-fic. The variety of pulsing frequencies also led to interference. In 1963, the Post Office thus abandoned research on analogue all-electronic exchanges. Economically, the equipment required was too expensive, while technically, it was clear that analogue pulse-amplitude modulation would continue to produce interference and incompatibility between different pulsed frequen-cies.[30] Flowers reportedly could not endure this decision and so resigned from the Post Office in 1964.[31] Highgate Wood, however, remained open, as the Post Office felt that "Highgate Wood has prestige value. . . . It will be the only electronic exchange really connected to the public system in Britain . . . and has been an important feature of the visits of parties from abroad. Arrange-ments are in hand for a visit of continental technical journalists in March [1964]."[32] The experimental Highgate Wood, disconnected from the tele-phone network, thus remained in place as a symbol of British prestige until the Post Office finally removed it in 1965. Highgate Wood lasted this long because both the warfare state and techno-nationalism sustained its develop-ment. While wartime relationships secured early support for Highgate Wood, the project's "world-first" goal as an export prospect was typical of Britain's techno-nationalist political economy. More importantly, however, both trends were tied to the shape that the British state took in the two decades following World War II, a shape built on war-forged relationships and inter-national standing amid a fragile empire. After Highgate Wood's failure, the Post Office and the government turned to new combinations of technocracy and managerialism in Britain's telecom infrastructure.

DIGITALIZATION AND CYBERNETICS

Flowers's departure from the Post Office paved the way for new digital approaches. In 1963, amid the failure of Highgate Wood, Flowers received a research report exploring an "integrated" digital telecommunications system, capable of transmitting both voice and data.[33] The report compared this integrated system with a specialized data network and concluded that specialized data networks might be preferable. The following year, Post Office researchers undertook another exploratory study shortly before Flowers resigned, and again, the integrated digital network concept received a muted reception.[34] After Flowers resigned, however, the new deputy engineer-in-chief, James Merriman, promoted that year, set up a "Rationalisation of the Distribution Network Working Party." This working party concluded in January 1966 that an integrated digital network, providing voice, data, and video, was possible.[35] During these trials, telecom engineers identified that more advanced cables could provide up to nine television channels, along with voice and data, although they recommended a lower bandwidth system, which would be cheaper.[36] Merriman, soon appointed senior director of engineering, the renamed engineer-in-chief position, ordered a follow-up study that confirmed this warmer reception by outlining the economic rationale for such a network.[37] After these reports, the Post Office board approved a trial for Washington New Town in 1967 as the first step toward a "single all-purpose cable to each home—an integrated network."[38] The Washington trial proved highly profitable, and further experiments began in Irvine, Craigavon, and Milton Keynes.[39]

Alongside digitizing transmission, engineers made good progress in digitizing telephone switching. After Highgate Wood, the Post Office dropped analogue pulse-amplitude modulation. It kept time-division multiplexing, allying it instead with digital pulse-code modulation, which the British radio engineer Alec Reeves first developed in the 1930s at International Telephone & Telegraph's Paris research center, Les Laboratoires Standard.[40] The Post Office began work on Empress, a prototype digital "tandem" exchange, which connected two local exchanges rather than providing direct connections to telephone subscribers. Empress's development went very smoothly, and in 1968 the Post Office installed Empress near Earl's Court, London, becoming the world's first operational digital telephone exchange.[41] The success of the integrated network and Empress trials, which both featured in Merriman's 1967 speech, vindicated the digital turn taken since Highgate Wood's failure and Flowers's departure. But while these successes help explain why and when Merriman announced the Post Office's vision of a universal digital infrastructure, they do not help explain the shape of this vision. In his speech, Merriman not only announced the digital network, but also that it would be general-purpose, self-healing, self-governing, and that it would unite information, control, and management. This represented a profound shift in thought, reconceiving the telephone network as an information network.

This vision's first influence was information theory and cybernetics, two connected fields studying information and control. In 1948, the Bell Labs mathematician Claude Shannon published "A Mathematical Theory of Communication," and the MIT mathematician Norbert Wiener published his book *Cybernetics*.[42] Both works contained remarkably similar theories of information, construing information as the amount of order or disorder in a selection of messages, although Shannon and Wiener differed a little in their specific definitions.[43] Shannon was inspired by his wartime work on statistics and cryptography, including the encrypted radiophone system SIGSALY, via which Roosevelt and Churchill talked. He thus took a tight, nonsemantic approach, in which information was a formal mathematical expression of the disorder, or unpredictability, of a message. Wiener, on the other hand, had worked on automatic antiaircraft weapon control systems. These control systems, which used radar feedback to track and predict planes' flight paths, served as a model for Wiener's theory of cybernetics. For Wiener, cybernetics was a theory of communication and control that explained the behavior of organisms and machines in terms of information inputs and outputs, just as the anti-aircraft predictors behaved based on information inputs and outputs from radar tracking. For Wiener, information theory was a corollary of cybernetics, used to analyze the information feedback loops in these systems, and so he took a semantic and pragmatic approach to information, looking at its meanings and effects on recipients. Wiener's novel science of cybernetics captured the public and scholarly imagination in the 1950s and 1960s, particularly in how it blurred the lines between machines and organisms and in Wiener's predictions of a "second industrial revolution," soon reframed as an "information revolution."[44] Meanwhile, information theory also became popular, leading to an "information bandwagon" in academia, where fields from communications to psychology became "informational."[45]

Post Office engineers paid close attention to Wiener and Shannon's ideas. In 1951, Roy Harris, from Flowers's Highgate Wood team, and D. L. Overheu, another Post Office engineer, attended lectures at Imperial College, London, to learn about information theory from Colin Cherry and Dennis Gabor. Along with Donald MacKay at King's College London, Cherry and Gabor made up the English School of information theory.[46] Harris and Overheu reported on information theory to Flowers and subsequently lectured on information theory to Post Office engineers around the country. At the end of their lectures, they thanked Colin Cherry for his correspondence on "information theory and cybernetics."[47] Harris and Overheu mainly addressed Shannon's "communication theory," but they also referenced Wiener's *Cybernetics* and his expansive definition of information, explaining how information theory could integrate subjects as diverse as digital computing and neurobiology by treating all forms of communications as the same.[48] They also used information theory to show how digital pulse-code modulation was less error-prone and delivered higher bandwidth than analogue pulse-amplitude modulation, foreshadowing its failure at Highgate Wood.

These reports expanded beyond Shannon's communication theory to reflect on the communications, biological, and industrial applications of information theory and cybernetics and on an impending "second industrial revolution," again quoting *Cybernetics*. Harris and Overheu pointed out how the versatility of information theory meant that it could assess the efficiency of transmission systems, switching systems, and even control organizations. They extended this further to consider cybernetic analogies between machines and organisms:

> Such machines are self-controlling and are given only a general instruction. Self-guiding projectiles, anti-aircraft predictors, etc., are examples of such machines. The human body is full of such mechanisms which control our temperature, balance, heart rate, etc., and there is evidence that conditioned reflexes operate in a similar way with a memory incorporated in the mechanism. Similarly the faculty of learning must depend considerably on past experience and must use some form of feedback.[49]

In this quote, Harris and Overheu advertised their cybernetic knowledge by referencing two classic examples of cybernetics, Wiener's antiaircraft predictors and the British cybernetician W. Ross Ashby's "homeostat," a device that modeled learning in the human brain by representing environmental adaptation through electrical feedback.[50] Harris and Overheu concluded by reflecting on Wiener's prediction of a "Second Industrial Revolution" from *Cybernetics* to hypothesize that this revolution could mean that "human beings, used as sources of judgement, may also be replaced by machines."[51]

These were expansive understandings of information theory that borrowed heavily from Wiener and cybernetics. By the end of the 1950s, three different definitions of information theory had emerged. The narrowest, associated with Shannon, referred purely to his communication theory. The second was a broader collation of Shannon and Wiener's work on analyzing communication problems. Finally, the third, which was particularly associated with the Imperial College information theory symposia and the journal *Information and Control*, was a synonym for cybernetics.[52] Post Office engineers followed this third definition, which was shown again in September 1952, when Post Office engineers attended another information theory symposium at Imperial.[53] They were joined by the English School and information theorists from the US, such as Stanford Goldman, Robert Fano and Yehoshua Bar-Hillel. One Post Office engineer, D. L. Richards, presented a paper on the effects of the physical properties of a circuit on telephone users' behavior, applying Wiener's pragmatic definition of information theory, viewing the communication system as composed of human and electrical components. A report on the symposium, delivered to Merriman, concluded that few papers at the symposium had "direct application to bread-and-butter communications practice," but suggested that information theory's main benefits may be "a fertilizing and catalytic action on the ideas of designers and development engineers, plus an understanding of ultimate possibilities that will act as a goal."[54]

By 1967, when Merriman was appointed senior director of engineering, cybernetics and information theory had indeed had a catalytic effect on engineers' goals for Britain's telecommunications infrastructure. The idea that the telephone network was an information network that could carry voice, video, and data was rooted in the homogenizing discourse of information theory. This was more than just rhetoric, as Harris and Overheu used information theory to demonstrate that digital transmission, the basis for this new network, was more efficient than analogue methods.[55] The natural extension of these insights was that the Post Office could use a digital telephone network for much more than telephony. Cybernetics too had a significant influence. Merriman built his vision of a "self-healing, self-governing" system on a cybernetic discourse that blurred organisms and machines, as did his two fundamental principles of "information and control," perhaps referencing *Information and Control*, the world's leading cybernetics journal. Merriman's vision of electronic computer control, wherein computers would receive information about traffic flow, determine the optimal route for calls, and control switching centers, leading to an autonomous, self-managing system, cybernetically blended organic and mechanical. This echoed Harris and Overheu's reports, which predicted that the "mechanised thinking of electronic brains together with self-stabilising servomechanisms" would allow machines "as sources of judgement" to manage industrial systems automatically.[56]

This quote also foreshadowed Merriman's seamless shifting between mechanical and managerial control, in which communications was "not only circuitry; it is man management."[57] Merriman explained to his audience that the Post Office's "greatest problem" was the "deployment, use, and, in one sense, control, of manpower."[58] Merriman outlined how the Post Office had increased productivity by harnessing "technological opportunity" and "scientific man-management" to found a new Management Services Department, which had undertaken studies on work measurement, queuing theory, and statistical control.[59] He used a diagram to demonstrate how the Post Office had conceived its works service organization as inputs and outputs of information (figure 1.1). He also outlined how the Post Office used computer-aided management techniques, such as Monte Carlo simulations, to improve engineers' workflows. These examples further suggest cybernetics' and information theory's influences, especially the works service organization as an informational feedback loop, which echoes Harris and Overheu's suggestion that information theory could assess organizational efficiency. But while cybernetics and information theory might explain some of these influences, they do not paint a complete picture or explain why Merriman emphasized administration and bureaucracy. The explanation for this lies with a British governmental tradition of bureaucratic mechanization, which was the second influence on the Post Office's digital vision.

THE GOVERNMENT MACHINE

Merriman had undertaken a similar approach to organization and management on secondment to the Treasury's Organisation and Methods department from 1956 to 1960, where he served as head of office machines.[60] Here,

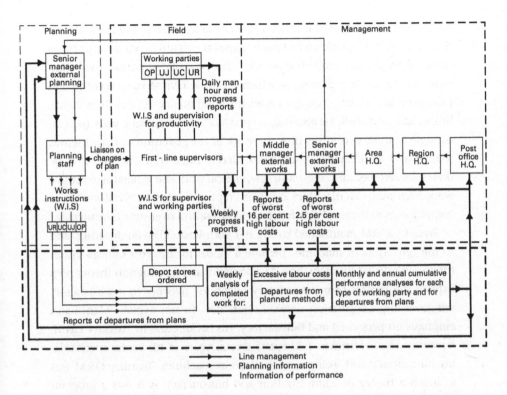

FIGURE 1.1

The Post Office works service organization as inputs and outputs of information. Source: Merriman, J. H. H. "Men, Circuits and Systems in Telecommunications." *Post Office Electrical Engineers' Journal* 60, no. 4 (1968): 241–251. Reproduced with permission from the Institute of Telecommunications Professionals.

Merriman was at the heart of Britain's "government machine," a tradition of discreet yet wide-ranging clerical mechanization that had run through British government since the nineteenth century.[61] In this tradition of "discreet modernism," expert mechanizers such as Merriman ran programs that mechanized and computerized clerical work, while carefully obscuring and setting apart this work from the executive work done by senior "generalist" civil servants. This discretion avoided the problematic notion that government work was wholly automatable.[62] Treasury Organisation and Methods (O&M), which had mechanized Civil Service payroll and statistics production, and computerized the Ministry of Pensions and National Insurance office, was this tradition's most important center. As head of office machines at O&M,

Merriman ran, in his own words, "a year of indoctrination and implementation" during 1959, in which he taught skeptical executive civil servants to see the Civil Service as a modern machine.[63] For example, Merriman used computer simulation "war games," in which executive civil servants competed in management simulations against academics and managers from Rolls Royce, British Rail, and Shell. He also staged "playlets" of bureaucratic work to demonstrate to these executives how the work of the government machine was routine and automatable. Merriman's success came from the line that he drew between specialist, mechanizable, clerical work and generalist executive work. Merriman cast machines as capable replacements for clerical work, and computers as managerial aids for executives' control over state bureaucracy.

Treasury O&M maintained strong ties with Post Office engineering, and so the "government machine" became a model for the Post Office's vision for Britain's digital infrastructure.[64] Cybernetics and information theory help explain the change from an ordinary telephone network to a self-healing, self-governing information network, but they do not explain Merriman's emphasis on personnel and bureaucracy. His background in Treasury O&M, however, does. Merriman had spent several years of his career mechanizing bureaucracy and seeing organizations as machines. Treasury O&M was as much a theory of administration and bureaucracy as it was a program of mechanization, a distinctive British administrative science that, given its prominence in the large-scale bureaucracy of government, was comparable to the management sciences of early twentieth-century America.[65] This explains why, for Merriman, any digital vision of Britain's telecom infrastructure had to combine administration with communication. Digitalization's universalizing tendency, in which everything becomes informational, and the government machine's compatibility with cybernetics and information theory, helped this mix.

Information theory, showing administration as informational flows, and cybernetics' vision of self-governing control systems, meshed well with the government machine as a science of bureaucracy and administration. At the 1958 Mechanization of Thought Processes cybernetics conference at Britain's National Physical Laboratory, one of the founding conferences for the new field of artificial intelligence, Merriman had presented a paper, "To What Extent Can Administration Be Mechanized?"[66] For Merriman, combining the government machine with cybernetic models of information and control was a natural basis for the Post Office's plans for a digital communications infrastructure, which explains why, in closing his paper, he asserted that, in this new digital infrastructure, "both men and systems become self-optimizing." This might not have been the first time that Britain's "government machine" influenced models of technology, as the British model of government bureaucracy quite possibly inspired Alan Turing's theoretical "universal machine," which in turn influenced the design of the modern stored-program computer.[67]

Merriman's emphasis on this infrastructure as a "general-purpose" network, however, was potentially transgressive. Governmental expert mechanizers had always been careful to delineate between specialist clerical work, which was mechanizable, and executive generalist work, which they avoided casting as mechanizable. This had roots in the longer history of the "government machine" metaphor for the Civil Service, which gained traction in the nineteenth century as a way to impart trustworthiness and mechanical objectivity to the new staff—women and lower-class men—drawn into the expanding Civil Service.[68] Casting women and lower-class clerks as mechanical components excused them from the upper echelons' expectations of gentlemanly secrecy and honor, while still affirming their reliability. Yet Merriman had no issue describing a "general-purpose" network that would automate management, and he directly opposed this vision against "special-purpose" networks. Merriman described the "considerable concern, and rightly so, at the possibility of a series of special-purpose networks evolving independently," which he contrasted with the "general-purpose digital network," which would have "versatility and open-endedness built in to permit becoming a better system in the future."[69] His emphasis on how this general-purpose network would computerize management appears taboo compared to the sacrosanct autonomy of the executive generalist civil servant. The explanation for Merriman's stance, however, lies with these "special-purpose" networks. The third influence on this vision was packet-switched data networks, which threatened the Post Office's monopoly.

COMPETITION AND CORPORATIZATION

From around 1960, data communication systems began to appear, particularly in the US. The US Air Force developed a system, the Semi-Automatic Ground Environment, or SAGE, that linked data from radar sites to warn of Soviet attacks on US airspace and began early operation in 1959. SAGE, in turn, influenced American Airlines' SABRE, or Semi-Automated Business Research Environment, which debuted in 1960 and was developed by IBM to book airline tickets automatically. Both systems worked over telephone lines and were thus the first "online" networks, proving the potential of networked data communications.[70] In the early 1960s, a new technique for data communications, packet switching, appeared. Paul Baran, a Polish American engineer at RAND, the American defense think tank, and Donald Davies, a Welsh mathematician at Britain's National Physical Laboratory (NPL), each independently developed packet switching in the early 1960s.[71] Packet switching divides a data message into small chunks, or "packets," which take separate, independent routes to the message's destination, where they are reassembled into the original message.

This differs from circuit switching, used by traditional telephone networks, which opens a continuous circuit along one route between sender and receiver, allowing the back-and-forth transmission of messages, as in a telephone conversation. Using packet switching, a destination can receive multiple messages from different senders simultaneously, whereas with circuit switching, the recipient can receive messages from only one sender at a time, occupying the line for the duration of that message. Packet switching was thus ideal for situations where multiple users needed to connect to the same computer simultaneously, a process called "time-sharing," whereas circuit switching was essential for voice calls. Baran and Davies's defense and scientific contexts shaped their work on packet switching. Hardening the American military communications network against nuclear strikes motivated Baran's research, while Davies developed packet switching to widen remote access to the NPL's scientific computing facilities. Time-sharing quickly commercialized, and the late 1960s was a boom period for the time-sharing industry. The two biggest companies in the industry, General Electric and Tymshare, built time-sharing computer networks that connected cities across the US. By 1972, Tymshare's network connected more than forty US cities and was expanding internationally.[72] This all brought increasing pressure within the UK for the Post Office to expand data services, and from 1967, a new group, the Real Time Club, began lobbying the British government to support packet switching and time-sharing.[73]

Merriman and the Post Office, however, were not so receptive, particularly to the Real Time Club, which complained about the Post Office's aversion to packet switching. In 1965, Davies sent a proposal to the Post Office to develop a national packet-switched data service jointly but received muted feedback.[74] While Merriman did not refer to packet switching explicitly in his 1966 presentation, he clearly stated that specialist data networks threatened the Post Office's goal of an integrated digital communications network based on Britain's existing circuit-switched telephone infrastructure. This attitude was even clearer several years later, in 1971, when Merriman warned the Post Office board that closed packet-switched data networks would infringe the Post Office's monopoly and obstruct its goal of a universal integrated network.[75] The Post Office has been called "ossified" for apparently missing packet switching's potential, and various histories of the internet have blamed the Post Office for Britain's lethargy in developing packet switching.[76] None of these histories, however, consider the alternatives that the Post Office might have had in mind. This was a time of diverse visions for data communications. In 1961, John McCarthy, the US computer scientist and AI pioneer, suggested that computing might become organized as a utility, like telephony, and Paul Baran lent credence to the computer utility proposal six years later, directly referencing the British Post Office's vision of a national data infrastructure.[77] As Merriman's speech shows, this vision extended beyond data to also include telephony, video calling, and television to the entire nation. There was doubtless a defensive, monopolizing quality to this vision, but from the Post Office's perspective, repurposing the national telephone network into a national information network was prudent and ambitious.

The Post Office's ambition, and its reorganization around a singular technological vision, stemmed from the fourth and final influence on the Post Office, which was its shift from a Civil Service department to a nationalized corporation. This shift began with the Post Office's financial troubles and a growing managerialist and technocratic spirit within British government. In 1957, Britain's balance-of-payments crisis and the government's stop-go economic policy cut public-sector investment, reducing the telephone business's ability to finance investment. As a result, the 1959 Conservative manifesto pledged to formally separate the Post Office's finances from the Treasury after a trial separation began in 1955. After the Conservatives' reelection, the 1961 Post Office Act revived a surplus trading fund for the Post Office, allowing the organization to balance its books. Various events and critiques throughout the 1960s highlighted that this act had neither loosened external controls as much as promised nor had the telephone business improved much.

Successive Treasury capital restrictions in 1962 and 1963 limited the Post Office's borrowing, and so the 1961 act was more a "modest milestone" in the Post Office's quest for freedom from the Treasury.[78] In 1963, the National Economic Development Council, founded by the Conservatives a year earlier to reverse Britain's poor economic performance, attempted to rectify this by announcing a near £900 million five-year spending plan for the telephone service.[79] In 1966, however, Labour prime minister Harold Wilson's "July measures," taken to avoid devaluing the pound, cut government spending, reducing the telephone business's investment program by £11.5 million.[80] The sluggish demand for telephone service during these restrictions motivated further critiques of the telephone business. The Brookings Institution, an influential US think tank, sharply criticized the 1961 act in its 1968 report, *Britain's Economic Prospects*, arguing that the act had unintentionally lowered telephone service demand by enabling the Post Office to raise tariffs, while various newspaper editorials demanded greater freedom and a more commercial attitude for the Post Office.[81]

Britain's rising public expenditure fueled technocratic managerialism in British government. Managerial modernization was in vogue in the UK in the late 1950s and early 1960s, emulating trends from the US. In 1956, William Whyte's *The Organization Man* highlighted and critiqued the growing numbers of US corporate managers, and management consultants such as McKinsey imported managerial thinking to the UK.[82] The government's 1961 Plowden Report on public expenditure, for example, made managerial sophistication its key target, and this report has since established a reputation for triggering a managerial revolution in British government.[83] This reputation is perhaps overstated. Computerized systems for management and expenditure planning were in place in the Treasury several years before the report, as the report both noted and praised. Only six years later, another review, the Fulton Committee, also recommended greater managerialism to solve rising public expenditure.[84] While the Plowden Report may have been neither especially novel nor successful, it nevertheless connected the time's managerial spirit with the public expenditure concerns that ailed the Post Office.

Technocratic managerialism had taken hold within the Post Office since the late 1950s as a way of negotiating these expenditure restrictions. The Post Office imported "accountable management" and "management-by-objectives" from the private sector and introduced computers as managerial tools.[85] In 1957, the Post Office opened its first computer center, the London Electronic Agency for Pay and Statistics, or LEAPS, with the promise that "the drudgery and, by modern standards, inefficiency of many dull, repetitive clerical routes will be swept away by the 'electronic office.'"[86] In 1965, a second London computer center followed in Charles House, Kensington, computerizing the preparation of customer statements and bills, which *Post Office Magazine* heralded with the headline "The Post Office Enters the Computer Age."[87] Highlighting this, the Charles House Computer Centre featured in a Post Office publicity poster series titled "Progress" (figure 1.2).[88] This computerized managerialism further reinforced the tradition of Civil Service bureaucratic mechanization within the Post Office. It also shows that the "white heat" of technocracy usually associated with Harold Wilson's first Labour government, elected in 1964, has a much longer history.[89]

Nevertheless, the Labour Party's election victory in 1964 was central to the Post Office's corporatization and its vision for a universal digital infrastructure. The new prime minister, Harold Wilson, appointed Tony Benn, a young modernizer, as postmaster general.[90] Benn's priorities were the "break," reforming the Post Office into a nationalized corporation, and the "split," dividing posts and telecoms into separate businesses. In March 1965, Benn argued to Wilson that Post Office reform was "a necessary act of modernisation," permitting the telephone business to become more entrepreneurial, dynamic, and effective in collaborating with the private sector.[91] The Post Office's struggles and Benn's appointment triggered several reports and committees, including two Joint Working Parties of Post Office, Treasury, and Cabinet officials; an National Economic Development Council subcommittee; an inquiry by the Select Committee for the Nationalised Industries; and a review of the Post Office by McKinsey, the US management consultants.

The first Joint Working Party, chaired by the Treasury, was ostensibly investigatory, making no recommendations, but its investigation still surfaced friction between the Post Office and Treasury. One Treasury official compared the Post Office's attitudes to that of "a Colonial nationalist movement anxious to get rid of the shackles of imperialist rule!"[92] The working party's final report arrived in July 1965 and was inconclusive, but perhaps unsurprisingly, given the Treasury's chairmanship, drew attention to the disadvantages of corporatization, suggesting that the Post Office was already akin to a nationalized industry.[93] James Callaghan, the chancellor of the Exchequer, was more explicit, expressing to Wilson his "doubt whether there

The Computer Centre in Kensington handles telephone billing for London subscribers, Savings Certificate repayments and other work. This and similar Centres will establish the Post Office as the leading commercial user of computers in Europe.

FIGURE 1.2
"Progress": the Kensington Computer Centre, 1965. Source: TCB 420/IRP (PR) 4, BT Archives. Courtesy of BT Group Archives.

is much profit in pursuing the matter just now."[94] Benn, however, countered with a historical treatise of the Post Office's struggles under Treasury control, which he presented to Wilson as "a hundred years of argument for reform."[95] The prime minister thus agreed to a second committee reviewing the structure of the Post Office, effectively siding with Benn on the break. The task of this second Joint Working Party was to frame the nature of reorganization, and it concluded that the critical question was the "split"—whether the Post Office should become one or two separate corporations.[96]

A National Economic Development Council subcommittee of cabinet ministers also considered the split. The committee studied Benn's arguments and the working parties' reports and, by February 1966, concluded that the Post Office should become a nationalized industry and public corporation, with one board supervising both the postal and telephone businesses, which should be separated at the executive level so that each could function independently.[97] The March 1966 general election, along with Benn's appointment as minister of technology, replaced by Edward Short as postmaster general in June 1966, delayed a public announcement. This delay effectively halted Benn's ambitions for the full separation of the postal and telephone businesses, and Short soon announced corporatization on August 3, 1966, leaving the question of internal structure for later.[98]

Two further reviews into the Post Office's internal organization show the role of technology and managerialism in its subsequent corporate structure. The first review, by the Select Committee on Nationalised Industries, began after Short's announcement and noted in February 1967, in its final report, that the Select Committee's conclusions would guide the Post Office's future organization. The committee offered several possible corporate structures for the Post Office. First, a federal structure with semiautonomous regional boards, comparable to gas and electricity in Britain and the Regional Bell Operating Companies of the US Bell System. Second, a semi-split, in which a coordinating board oversaw separate, largely autonomous executives for posts and telecoms. Third, and favored by the committee, re-creating the Post Office as a corporation, maintaining central support services.[99] The committee argued that telecommunications' imaginative, managerial technocracy would inspire technical thinking in the postal business.[100] The Select Committee considered the postal service's loss of the Post Office research station at Dollis Hill in north London, which mainly undertook telecommunications research (discussed further in chapter 5), as "the greatest drawback" of separating the postal and telephone services.[101]

While the Select Committee's report influenced the Post Office's eventual corporate structure, the McKinsey report was more influential on the Post Office's digital vision. Ronald German, the Post Office's director-general, had hired McKinsey in 1965 to review the postal business, but Benn quickly embraced McKinsey as a potential ally in the fight for the split.[102] McKinsey's appointment further indicates this period's managerialist spirit. McKinsey's reputation grew so quickly in Britain during the 1960s that, by 1969, a journalist for the *Times* could write, "Ask anyone to name a management consultant and chances are, if he is British, that the answer will be 'McKinsey.'"[103] McKinsey recommended increasing the independence of the telephone and postal businesses and enhancing the telephone business's control over engineering and research by abolishing the shared Engineering Department and moving research to telecommunications' control.[104] McKinsey also concluded that telecom development needed more "technologically complete" plans, over a thirty-year timescale. This recommendation soon manifested in the plan for a universal digital integrated network, the most expansive and comprehensive plan for British telecommunications formulated since World War II. This recommendation also led to the Post Office creating a long-range planning department to sustain that vision into the 1980s (discussed in chapter 2), as well as specific reports, committees, and projects to apply that vision to telecom switching and transmission (discussed in chapters 3 and 4).

The Post Office's corporatization, and the associated recommendations for more holistic and managerial technological plans, thus shaped the vision for a universal digital infrastructure for the UK. Not all these recommendations were followed to the letter, however. While the government approved the break, as proposed by the second Joint Working Party and National Economic Development Council subcommittee, it rejected the split or semi-split, as had been advocated by Benn and McKinsey, after the Select Committee argued that central functions, like research, strengthened the case for keeping the postal and telephone businesses together.[105] The Post Office board thus followed McKinsey's recommendations in other ways, appointing separate managing directors for the postal and telephone businesses; abolishing the Engineering Department and position of engineer-in-chief, so that Merriman instead became senior director for engineering; and moving research to the telephone business. The Post Office established separate headquarters for the postal and telecom businesses, while a central headquarters housed support functions. By October 1, 1969, when the Post Office Corporation was formally vested, this new corporation had thus devoted nearly all research, planning, and engineering resources to tele-communications to support the vision for a universal digital infrastructure.

CONCLUSION

The digital vision emerged from the analogue failure of Highgate Wood, which showed the lingering influence of the warfare state on the Post Office. The Post Office's wartime work at Bletchley Park had lent gravitas to Highgate Wood and had forged the personal relationship between Tommy Flowers, Highgate Wood's chief designer, and Gordon Radley, controller of research and later director-general, that sustained the project. Moreover, the development of Highgate Wood shows how a nationalist political economy affected the Post Office at the end of the 1950s and in the early 1960s. Its focus on Highgate Wood came from the government's ambition to promote Britain on the world stage and secure export prospects for British electronics. The failure of Highgate Wood thus meant change on several fronts. The Post Office prioritized national prestige and export prospects less, which meant focusing less on a singular exportable product and more on a systematic future for Britain's telecom infrastructure. Finally, personnel changed, as Tommy Flowers and his Bletchley Park legacy departed, and two new engineers—Roy Harris, Flowers's deputy, and James Merriman, first as deputy engineer-in-chief and then as senior director of engineering—came to the fore.

Harris and Merriman's past experiences had already sown the seeds of this new vision. Harris was one of the Post Office engineers most interested in cybernetics and information theory, attending the influential London Symposia on Information Theory and lecturing Post Office engineers around the country on cybernetics. Meanwhile, as the Treasury's head of office machines, Merriman had become one of the most influential mechanizers in Britain's "government machine." Together, cybernetics, information theory, and the government machine composed the intellectual foundation for this vision. Drawing on cybernetics and information theory meant seeing the telephone network as an information network that could use feedback loops and decentralized machine control to become self-healing and self-governing. Furthermore, cybernetics set the expectation for a forthcoming information revolution, which was particularly influential on the Post Office's practices (as detailed in several of the following chapters). The government machine lent to this framework an ambition to include both bureaucracy and communications, using feedback loops and machine control to manage labor.

Alongside these ideas, two other pragmatic influences shaped this vision. Packet switching brought the threat of independent data networks, incentivizing an integrated digital vision that extended the Post Office's monopoly from telephony and telegraphy to data and video services. The Post Office's corporatization brought managerial and structural changes. Both governmental reviews, such as the 1961 Plowden Report, and key politicians, such as Tony Benn, saw technocratic, corporate managerialism as the solution to the financial and organizational failures of both government and the Post Office. This ethos created a hospitable environment for a vision that was as much about administration as it was about information. This ethos also fueled reviews into the Post Office, leading to its corporatization and greater independence. The McKinsey review, advising that the Post Office develop long-term, technologically complete plans and reorganize engineering and research to devote more support to telecommunications, appears particularly influential. These reviews all meant that, by 1966, Merriman had articulated this long-term, holistic vision, and the government had announced the impending corporatization of the Post Office. By 1969, the Post Office had become an independent nationalized corporation and reorganized its internal technical resources to support this vision.

Thus, by the end of the 1960s, an ambitious vision was in place to launch a national digital infrastructure. This vision, of an automated infrastructure, responsive to perceived threats and crises, such as competition, was more than just a technological vision. It also had an explicit view on labor relations, seeing digitalization as more "man management" than circuitry. The move to embed computer control throughout communications and bureaucracy in this infrastructure would have troubling implications for both staff and customers. Personal networks, especially those that connected Post Office engineering and Treasury O&M, supported this vision, and the Post Office's corporatization meant that material and managerial structures were already in place to implement this vision. Corporatization was key. This vision was also the product of the Post Office gaining more independence from government oversight. In doing so, the Post Office became more like a private corporation, no longer led by an elected official in the postmaster general, but instead by a chairmen and managing directors. Furthermore, this vision was the product of a monopolizing mentality, in which all communications came under the natural purview of the Post Office and consequently the Post Office needed to commercialize digitalization before specialist, private-sector competitors. This mentality even eclipsed packet switching, which had come not from the private sector but from the National Physical Laboratory, another state institution. Merriman's 1966 speech thus bore the traces of the many influences on this vision, showing its depth and origins. The next three chapters, in turn, show the breadth of this vision and its effects, demonstrating why this vision mattered to the digitalization and privatization of British telecommunications.

2 PLANNING THE DIGITAL FUTURE: MODELS AND MARKETS

In 1966, the Post Office founded the Long Range Systems Planning Unit to expand on its vision of a national digital infrastructure and to develop plans that would turn this vision into reality.[1] One of the key figures in this unit was Roy Harris, who had worked for Tommy Flowers on the Highgate Wood project and had reported and lectured on cybernetics and information theory for the Post Office. In keeping with the Post Office's ambitions for long-range planning, Harris's mission statement declared that the "prime purpose of planning was to invent the future, not to predict it."[2] Long-range planning, however, was not characterized purely by limitless techno-optimism for this digital future, and planners also pondered the problems of a self-governing, computer-controlled national infrastructure. In 1969, J. S. Whyte, the head of long-range planning, warned about the "bleak mechanistic prospect" of computerized telecom infrastructure, arguing that planners should mind the risk that such an infrastructure could erode dignity and privacy.[3]

By 1980, long-range planning looked very different. Preparing for the end of monopoly, BT renamed the unit the Business Planning and Strategy Department. Rather than "inventing" the future, the department made extensive use of computer models that would "scan" the future, while "competition and diversity" were proclaimed as "ideas of the future."[4] Computers would help realize this future too, as planners and executives discussed how the digital convergence of computing and communication could aid "small government."[5] Moreover, what Whyte had labeled bleak prospects for computer-controlled infrastructure in 1969 had turned into opportunities by 1980. At a long-range planning seminar with BT's senior management, planners and executives anticipated how predictive computer control could allow autonomous infrastructure to anticipate users' information habits.[6] Combining this with information filtering, they speculated, could enable BT to "mould society." Several changes had thus happened to the Post Office's plans for a national digital infrastructure between 1966 and 1980. They had turned from inventive planning to predictive modeling, from warning about privacy to eroding it, and from emphasizing public monopoly to promoting small government and corporate surveillance.

This chapter continues the focus on the Post Office's visions and expectations for Britain's digital infrastructure. As chapter 1 explored how state ownership was central to the birth of this vision, so this chapter explores how that vision grew during national ownership, and how privatization and liberalization changed it. But it moves beyond a history of visions and expectations to explore the powerful role of explicit future-making practices like planning, forecasting, and futurology. The history of planning is central to understanding British political economy in the 1960s, when planning's appeal surged because it offered several opportunities. Planning could maintain the government's popularity with a public that increasingly perceived government's role as securing faster economic growth. It also offered a way to keep inflation and unemployment low and a solution to the demands of Britain's growing population on the welfare state. Finally, it tapped into the broader technocratic mood of the time. Britain's plans included macroeconomic, public expenditure, healthcare, and regional and housing planning. Ultimately, however, these plans overpromised and underdelivered, creating public disillusionment in the short term and, in the long term, paving the way for neoliberal politics that rejected both planning and state intervention.[7]

It would be a mistake, however, to see the Post Office's creation of a long-range planning unit as a last gasp of Britain's planning moment. Britain's 1960s planning vogue had roots in statistical forecasting and industrial interventionism from the interwar period, and while much of this activity focused on the state's capacity to create "long-term" plans, "long-range planning" was instead part of a technocratic response to planning's failures in the 1960s.[8] Long-range planning and other futurological fields were political "prediction technologies" intended not to forecast statistical, economic futures but instead to promote freedom of choice over optimal futures, embedding liberal politics in futurology. Yet, technocratic planners often remained as gatekeepers to these prediction technologies and their futures.

This was a controversial temporal extension of technocracy, particularly as many of these techniques emerged from the US defense think tank RAND. Public figures from Hannah Arendt to Lewis Mumford thus protested prediction's emergence as a technocratic Cold War conquest of the future.[9] This chapter thus maintains focus on the technocrats at the heart of Post Office telecom engineering who developed and modified their prediction techniques in dialogue with Britain's shifting political economy.

Futurology and futures research also paved the way for neoliberal imaginaries.[10] In this period, prediction turned from societal choices about world futures to paid, managerial consultancy work on policy and corporate decision-making. Focusing only on this end point, however, would miss that these prediction technologies were not born neoliberal, but instead developed through the influence of the US state-military-industrial complex, as well as French dirigiste planning circles. Prediction technology could thus "bridge statist and corporate rationalities and bring management methods taken from the large corporation to bear on governmental ways of 'seeing the future.'"[11] This chapter explores prediction technologies within British state institutions, the Post Office and BT, as they developed alternative modes of technocratic future-making in response to planning's failures, the corporatization of the Post Office, and the creation of BT. Studying these material prediction practices, whether they were complex computer models or simple timelines and budgeting documents, is vital. They materialized specific futures, visions, and expectations and sustained the communities and technologies invested and embedded in those futures.[12] The Post Office and BT mobilized these prediction technologies, which mainly took shape as computer modeling and simulation, to defend the digital vision through the 1970s and 1980s. Long-range planners found digital tools that would help them plan and implement a digital infrastructure, tools that both contested and adapted to the market turn and, in doing so, transformed the Post Office's plans for Britain's digital infrastructure from monopoly to market, from discreet to invasive, from public to private.

INVENTING THE FUTURE

In 1966, James Merriman set up the Long Range Systems Planning Unit to advise telecom R&D.[13] Its creation came after the McKinsey review of the Post Office, which had suggested that the Post Office needed more effective

long-term direction of R&D through "commercially sound" and "technologically complete" plans over a thirty-year timescale.[14] The Highgate Wood
failure, for which the Post Office had attracted criticism in the *Sunday Times*
and from the Select Committee on Nationalised Industries, also influenced
the department's creation. The Post Office's director-general, Ronald German, answered the Select Committee's critiques of Highgate Wood's failure
by pointing out that a new group, the Long Range Systems Planning Unit,
had been formed to look at the "broad brush" of technological development.[15] Roy Harris, formerly of the Highgate Wood team, joined the Long
Range Systems Planning Unit, and in this position, he penned two documents that formalized the Post Office's vision of a universal, integrated digital network. These two reports were "Telecommunications System of the
Future," the department's founding document, and its sequel, "Telecommunications Systems of the 1980s."[16]

In "Telecommunications System of the Future," Harris advocated an
"inventive" approach to the future. His report outlined a series of technologies, from videophones to remote computing, for which the department
would coordinate both research and provision. Long-range planning would
"exploit" these possibilities while also allowing for unexpected demands
from telephone service users. Harris's sentiment—that "the prime purpose of planning was to invent the future, not to predict it"—captured his
assumptions about R&D's limitless inventive capacity.[17] This turn of phrase
echoed Hungarian-British electrical engineer Dennis Gabor's expression
that "the future cannot be predicted, but futures can be invented," which
appeared in his 1963 popular science book, *Inventing the Future*.[18] Harris
may have borrowed this phrase after a personal encounter, as both Harris
and Gabor attended information theory symposia at Imperial College, London, and Gabor later collaborated with Post Office researchers on computer
simulation of speech compression.[19]

For Harris, this inventive approach to the future revolved around plans for
the integrated digital network in two ways. First, it meant reinforcing expectations about social and technological change that demanded an integrated
network, and second, it meant picking the systems that would support that
network. "Telecommunications System of the Future" and "Telecommunications Systems of the 1980s" therefore both made broad claims about the future
relationship between information technology and society and outlined the
general principles of integration, digitalization, and computerization for the

universal digital network. In "Telecommunications System of the Future," Harris anticipated growing demand not just for telephone services but for a whole range of information services, including TV and radio broadcast, Viewphone facilities, fax transmission, remote computing, information retrieval, and data processing services. And, as Merriman had declared in 1966, the key to the Post Office's future was integrating telecommunications through computerization and digitalization so that the network could deliver all these services. Harris outlined the "total system concepts" that would allow the Post Office to meet all these demands, the first of which were "integration" and "information concepts." Harris spoke about how it was "essential to adopt an integrated approach whereby an increasing variety of telecommunication services is provided by the same general-purpose telecommunications network." He emphasized again how information theory lent the general concepts and techniques necessary to run these diverse services across a single general-purpose network. Harris then discussed the importance of "general-purpose control," in which, if the Post Office were to run a digital network providing integrated information services, applying computerization to network management would be essential.

In "Telecommunications Systems of the 1980s," Harris expanded on these points and made recommendations that dictated some of the Post Office's largest projects in the 1970s. Again, Harris repeated the Systems Planning Unit's expectations that the society of the 1980s would demand more information services and depend more on them. Integration was the ideal solution, and Harris referred to the integrated network field trials happening in new towns such as Washington and Milton Keynes. Harris then made a series of recommendations for Post Office engineering, all of which the Post Office followed. He recommended a forecast of the future distribution of different types of telephone exchange, which was one of long-range planning's most important projects and would later become a national controversy. Harris also recommended a broad, long-term review of the Post Office's long-distance network and how it could best transmit information services. This was another important project for long-range planning that (as discussed in chapter 4) fueled both the development of integrated digital standards and another high-profile technological failure in the Post Office. He emphasized the importance of studying local users' needs and when they would need various information services, such as video calling, which became another focus for long-range planning through the 1970s. Finally, Harris recommended that the Post Office undertake a six-month feasibility study with industry on computerized telephone exchanges and follow that study with a five-year R&D program.

This final recommendation ended Harris's tenure at long-range planning. In the summer of 1967, Harris moved to become head of telephone exchange development and chair of the Advisory Group on Systems Definition, set up with industry to establish the technical feasibility and specifications for the digitalization and computerization of telephone exchanges.[20] As part of the Post Office's corporatization and reorganization, the Long Range Systems Planning Unit also underwent several changes. Long-range planning changed from a unit advising R&D to a fully-fledged division advising the telecom side of the Post Office. In doing so, the newly named Long Range Studies Division also gained a new head. In 1969, J. S. Whyte returned to the Post Office from secondment to a familiar destination, the Treasury's O&M department. As Merriman had done ten years earlier, Whyte had spent most of his time working on state computing projects and had helped restructure the British computing industry into ICL, a new flagship manufacturer that the Labour government hoped would be internationally competitive.[21] Whyte's return to the Post Office as department head for long-range planning, reporting directly to Merriman, the new senior director for engineering and another Treasury O&M alumnus, perhaps suggests that Merriman was stacking senior positions with like-minded engineers. Whyte committed to Harris and Merriman's inventive futures of integrated digital networks and spread bureaucratic computerization throughout long-range planning. Unlike the discreet mechanizers in central government, however, Whyte would take a more visible role in reaffirming the Post Office's role as the steward of national communications.

COMPUTER MODELS AND MECHANISTIC PROSPECTS

Under Whyte, long-range planners remained committed to inventing digital futures. This commitment is perhaps most evident with Viewphone, a videophone terminal that appeared in Harris's reports, Post Office promotional films, and further plans for the integrated digital network.[22] Viewphone was a desktop terminal with a seven-by-five-and-a-half-inch screen and a loudspeaker-equipped telephone for two-way speech and vision. A long-range planning report on Viewphone declared that, with rising income, growing business advantages, and the "continuing advance in technological capability," it was "virtually certain that demand for a viewphone service will arise in the future."[23] AT&T's Picturephone, launched in 1964 to great fanfare but ultimately a commercial failure, heavily influenced these plans. Long-range planners concluded that Picturephone failed only because the public was not ready for video telephony and advised that the development of a British videophone should proceed because of the inevitable demand for the Post Office's integrated information network. Internal trials started in 1972, linking Telecommunications headquarters in central London with the Post Office's research station at Dollis Hill, northwest London.

Viewphone ultimately failed, but in doing so, it demonstrated long-range planning's success. A 1976 report on the Viewphone trials revealed little enthusiasm, as users found that many calls were blurry and that visual contact did not improve communication.[24] Users also mainly employed Viewphone to transmit written material, neglecting video telephony itself. Although BT still showcased Viewphone throughout the 1980s (a 1988 *BT Journal* article described it as an "important new service"), there was no commercial development until 1993, when BT launched Relate 2000, its first videophone.[25] This history echoes AT&T's Picturephone, which, despite its failure, was "a rather successful piece of the technological imagination that guided innovators by helping to establish a basic paradigm for information services and technology."[26] Even after unsuccessful trials, Viewphone's continuing presence demonstrates the Post Office's commitment to a single national network for voice, data, and vision.

The department also began using computer modeling for two of Harris's recommendations from "Telecommunications Systems of the 1980s." The first use was to forecast the long-term development of Britain's long-distance transmission network, and the second was to forecast the rollout of new telephone exchanges. The first project came together through the creation of a new group, the UK Trunk Task Force, within long-range planning. This group's task was to recommend a strategy for trunk network development up to 1985, with looser recommendations until 2000. The task force modeled traffic across the entire network for telephony, data, and, of course, Viewphone. The model simulated this traffic's cost, quantity, and quality based on whether transmission, signaling, and switching were analogue or digital. The model also simulated different network layouts to find the best arrangement of switching centers for the network's "backbone."[27] The model found that digitalization would halve equipment costs and predicted that telephone connections would double by the decade's end, nearly quadruple by 2000, and that total traffic would almost quintuple.[28] The UKTTF's report, completed in 1971, thus recommended the digitalization of the entire network and was endorsed by the Managing Director's Committee for Telecommunications in 1972.[29]

The task force's model was not a neutral calculation, and it contained several assumptions that betrayed its human inputs. The model assumed a future of widespread videophone use, meaning that the network would need to carry video, in addition to voice and data, and assumed this would culminate with Viewphone's transformation into a "concept that included facsimile and visual access to data banks."[30] The head of the task force, Denis Breary, admitted that "a certain amount of forecasting of a sociological nature was necessary to establish a likely pattern of demand in the latter decades of this century."[31] The model was thus working from the task force's assumptions, which included expectations that, by 1980, Viewphone and teleconferencing systems would be widespread and that, by 2001, digital transmission between local users and telephone exchanges would need to carry both telephony and Viewphone signals.[32] These assumptions were all based on the Post Office's goal to build an integrated digital network and the expectation that the society of the 1980s would require voice, video, and data services. These predictions were not outputs of the task force's model but rather inputs, suggesting that senior Post Office management endorsed the task force's report not because of the strength of its findings, but because it reinforced their existing visions and expectations for the future.

Computer modeling also forecast the rollout of new telephone exchanges. One of Whyte's first projects as head of long-range planning was to simulate the optimal depreciation and replacement rates for outdated electromechanical telephone exchanges over thirty years.[33] Every simulation strategy found that installing a hypothetical new electronic exchange would be cost-effective, and faster strategies were better than slower ones. Unfortunately, few details remain about this model's technical composition. Surviving records show that the model was called A Local Exchange Model (ALEM), a version of it ran on a Honeywell computer hosted by the Post Office's Management Services Department, and that it used discounted cash flow analysis to calculate the year-on-year costs of exchange replacement.[34] Finally, and crucially, the model assumed that the hypothetical replacement electronic exchanges in these simulations had the exact same cost and traffic capacity as quoted to the Post Office by the US-owned manufacturer, STC, for its new TXE4 series of electronic exchanges. In January 1971, as it would do the following year with the UKTTF report, the Managing Director's Committee for Telecommunications endorsed the simulations' conclusions. In April, the Post Office board followed suit and "thanked Mr. Whyte for his clear and comprehensive presentation of the telecommunications business proposals for exchange equipment," approving the purchase of STC's TXE4 based on these simulations.[35] Whyte handed the model over to the Operational Programming Department and liaised further to plan the TXE4 modernization strategy. Later that year, Whyte left long-range planning to become the director of operational programming.

The model's assumption that TXE4 would replace electromechanical exchanges shows a continuing commitment to invented futures. Whyte's model did not offer a new future but upheld a decision to roll out TXE4, thus supporting a future where the Post Office board purchased electronic exchanges from STC. There are parallels here with Viewphone and the trunk task force model. In all these cases, long-range planners picked a winner and then studied different ways to develop and roll it out. The department thus invented rather than predicted futures. As the next two chapters will show, however, this strategy could backfire. Using TXE4 as the basis for the ALEM model would contribute to one of the Post Office's most controversial procurement decisions of the 1970s. The trunk task force's model and its expectations about Britain's need for a high-bandwidth telecommunications infrastructure also guided investment in the waveguide, an experimental transmission technology that ended up as a high-profile and costly failure. For now, however, this chapter will remain with the Long Range Studies Division and how it maintained and transformed the Post Office's visions and expectations.

One new area that Whyte focused on while head of long-range planning was public concerns about computing and digitalization. In various publications and events, he conceded that rapid technological development risked machine control over society. He argued that "machines must not be permitted to erode the dignity of man," warning about the "serious questions of the invasion of privacy" and titled one paper "Telecommunications in the Service of Man," inviting the question of what the alternative might be.[36] Whyte most forcefully articulated these concerns at a 1969 conference, "City in the Year 2000." The conference included eminent speakers such as Ray Pahl, the sociologist of suburban and postindustrial communities; Alexander Macara, the future chair of the British Medical Association; John Dennis Carthy, a prominent BBC science communicator; and Meredith Thring, fuel scientist, mechanical engineer, and future coauthor (with Eric Laithwaite, who created maglev transportation) of the 1977 popular science book *How to Invent.*[37] Whyte described how automated, computerized communication networks could degrade humankind:

> There seems to be no reason in principle why we should not envisage the fully automated situation in which the individual need rarely leave his home but merely manipulates the knobs and dials and screens around him in order to obtain his education, conduct his business, do his shopping and get his entertainment. This bleak mechanistic prospect is unacceptable because it pays no regard to the fundamental nature of man, and his indispensable need to interact with other men and seek self-fulfillment. . . . If men are to have any hope of controlling their own destiny, they must attempt to reduce the gap between our explosively growing technological capability and our lack of understanding of its social consequences.[38]

In warning about this "bleak mechanistic prospect," Whyte responded to public concerns about the dehumanizing intrusion of computers and communications into personal life. In 1967, Alan Westin, Professor of Law at Columbia University, had published his influential book *Privacy and Freedom.* This book drew attention to new, often technological, ways of invading privacy, such as the informational surveillance made possible by computerized data banks. The Younger Committee, Britain's first large-scale official study of privacy, thus invited him to present evidence, and its 1972 report underlined the threats of mass communication and computerized record-keeping systems.[39] This could be seen in the public hostility to the 1971 census and the uncertainty around public data protection, which led to the highest-ever proportion of the public refusing to complete the census. Three years later, Harold Wilson's new Labour government published two white papers, *Computers and Privacy* and *Computers: Safeguards for Privacy.* In this climate, British government computing projects had become interpreted as a new threat posed by the centralized, computerized state to the individual citizen.[40]

Whyte and Merriman came from a British governmental tradition of "discreet modernism," where expert mechanizers obscured their extensive automation and computerization of the British state. Whyte's visible warnings appear to defy this tradition, as well as undermine the Post Office's plans for digitalizing and computerizing Britain's telecom infrastructure. As head of long-range planning, however, Whyte no longer worked as an expert government mechanizer in Whitehall. Instead, he was part of a public monopoly that, as Harris and Merriman established in 1967, invented futures of computerized, integrated networks for telephony, data, and Viewphone. Discretion may have sufficed for Treasury O&Ms' expert mechanizers, but in the national telephone system, at a time of rising privacy concerns about computerization and telecommunications, Whyte could not be discreet. He thus mollified public concerns about computers and privacy by affirming the Post Office's awareness of the potential dystopian consequences of its plans, as well as its commitment to avoiding those consequences. This was also an important sign that, for senior Post Office engineers, a nationalized political economy for digital infrastructure was no longer infallible. They could not assume that nationalized control over communications was uncontroversial, and instead had to directly address the political dimensions of their project. At this point, under national ownership, computer control and surveillance were a threat to address. Throughout the 1970s and 1980s, however, as the telecommunications monopoly ended, and computer simulation became further entrenched within long-range planning, computerized control and surveillance would begin to look less dystopian to telecom engineers and managers.

SYSTEM DYNAMICS AND POSTINDUSTRIAL SOCIETY

In the 1970s, long-range planners began to study new uses and visions for computing inside and outside telecommunications. These studies emerged from several influences that changed long-range planning in the Post Office. The Post Office restructured the telecom business, moving long-range planning to a new department, Telecommunications System Strategy, and renaming long-range planning as the Long Range Intelligence Division. The new Telecommunications System Strategy Department, headed by Roy Harris, was created to oversee the development of the general-purpose network, particularly System X, Britain's first fully computerized telephone exchange. Although this move might seem to shrink long-range planning's responsibilities, relocation gave long-range planning a new focus on broader social and economic futures. Whyte, who left long-range planning in 1971, was replaced by Alex Reid, former director of University College London's Communications Studies Group, where he had directed research into the social impact of computers and communications under contract for the Post Office and the Civil Service.[41] Under Reid's tenure, long-range planning began social forecasting and hired researchers from across a wider range of fields, including information scientists, statisticians, sociologists, and psychologists.[42] The department's relocation and instigation of economic and social forecasting, along with Reid's appointment and new hiring strategy, suggest that Merriman and Harris felt that long-range planning needed a new direction, oriented more toward the telecommunication system's social and economic environment.

These influences, alongside the Post Office's strained economic environment in the 1970s, manifested in the quantity and quality of reports that long-range planners wrote from 1974, when the department relocated to Telecommunications System Strategy. Of the thirty-seven reports written before 1974, thirty-two addressed technological change in telecommunications, while five addressed societal futures.[43] Of these five, three profiled long-range planning in British government; one reported on the conference "City in the Year 2000," where Whyte had warned about the "bleak mechanistic prospect"; and the last, a 1971 study titled *Britain 2001 AD*, projected Britain's economic environment to the year 2001.[44] This last report was also the first recommendation for comprehensive economic forecasting in the Long Range Intelligence Division, citing the pressures of Britain's turbulent economic environment. In 1974, the department thus began two new series, Long Range Economic Forecasts and Long Range Social Forecasts. From this point, approximately 40 percent of long-range planning reports were economic or social forecasts, 25 percent were telecommunications forecasts, and 35 percent were "interactions" forecasts, synthesizing research on telecommunications futures with social and economic forecasts.[45]

These economic and social forecasts provided avenues for new uses and visions of computers in telecom infrastructure. The first Long Range Economic Forecast, "The Economic Consequences of Energy Scarcity," introduced the Post Office to a new type of computer modeling, system dynamics.[46] The report responded to Britain's 1973–1974 energy crisis, caused by a National Union of Mineworkers strike that slowed down domestic production of coal, and the 1973 oil embargo by OAPEC, the consortium of oil-exporting Arab nations. Prime Minister Edward Heath thus introduced a three-day workweek in December 1973 to conserve energy. The combined energy and economic crises caused telephone growth to drop by 50 percent, and so the board initiated a Telecommunication Energy Conservation Program, while long-range planners ambitiously studied the potential for telephone exchanges powered by on-site nuclear reactors.[47] The energy scarcity report surveyed and synthesized a range of forecasts from think tanks and policy units, such as the University of Sussex's Science Policy Research Unit and appraised the likelihood of future energy crises.

The report highlighted system dynamics' use in *Limits to Growth*, an influential 1972 report published by the Club of Rome, an intellectual network formed in 1968 to draw attention to issues requiring global action.[48] The *Limits* report used Jay Forrester's system dynamics, originally known as "industrial dynamics," which Forrester developed while at MIT's School of Industrial Management as a heuristic tool to model industrial systems and help managers better understand the systems they managed.[49] Industrial dynamics expanded into system dynamics, which, for the *Limits* study, became "world dynamics." Using world dynamics, *Limits* projected an "overshoot and collapse" of society based on the interaction of five variables: world population, industrialization, pollution, food production, and resource use. Post Office long-range planners dismissed this gloomy prediction but nevertheless concluded that a long-term energy problem was likely. They thus emphasized that, to weather future crises, business and government needed "more sophisticated" long-term planning.[50] Forrester's system dynamics would become a key part of the department's "more sophisticated" long-term planning.

Meanwhile, the department's social forecasts offered new ideas about computers' role in society. Joan Glover, the department's newly hired sociologist, undertook and analyzed customer interviews and questionnaires to forecast changes in labor structure, home working, and telecommuting. Glover concluded that networked computing would facilitate home working for professional, managerial, and clerical workers and transform work from the type where "people and machines were coordinated to produce goods" into the "co-ordination of people and machines to produce knowledge."[51] Expansive and academic, Glover's social forecasts cited Max Weber, Michael Young, Anthony Giddens, Peter Hall, Georges Friedmann, and Peter Berger on the nature of work, family, leisure, cities, and alienation. They also showed the continuing influence of US futurology, citing reports from the Institute for the Future, a RAND spin-off, and *The Year 2000: A*

Framework for Speculation on the Next Thirty Years, the highly influential futurological text by Herman Kahn and Anthony Wiener.[52] Perhaps the strongest influence, however, was Daniel Bell's *The Coming of Post-Industrial Society: A Venture in Social Forecasting*, published in 1973.[53]

Bell's *The Coming of Post-Industrial Society* combined computers and futurity in two ways. First, Bell predicted a future society in which information, rather than matter, would be the primary resource for economic growth. Computers were central to Bell's predictions, rendering work informational rather than material, and so Glover's futures of informational, computerized work, in contrast to pasts of material production, echoed Bell's postindustrial society. *The Coming of Post-Industrial Society* was a turning point for understanding how computers would change the future, and postindustrialism proved a popular concept to explain the 1970s zeitgeist that "computers soon seemed *everywhere*."[54] Second, *The Coming of Post-Industrial Society* was not only about capitalism's future but was also a plea for more extensive use of social forecasting techniques.[55] Bell argued that industrial society's linear forecasting could not understand postindustrial society, so society needed new predictive techniques to judge the various trends, indicators, and technologies that would shape the future. Computers, as "intellectual technologies," were again crucial. Using techniques such as simulation and model construction, these intellectual technologies would forecast and solve the problems of postindustrial society. Bell thus argued that computers, as an intellectual technology, would play a double role, both forecasting society's postindustrial future and, within that future, making society postindustrial by rendering work informational rather than material. This double role of the computer played a similar role in long-range planning as in Bell's postindustrial society. While long-range planners did not extensively reference Bell's work, his strong influence in Glover's social forecasts on postindustrial society suggest that Bell's arguments about computers as intellectual and prediction technologies affected long-range planners.

This influence is suggested by long-range planners' growing use of system dynamics simulations, one of futurology's archetypal intellectual technologies. The department first used system dynamics to respond to the potential reorganization of the Post Office's telecommunications monopoly. The economic constraints imposed on the Post Office during the first half of the 1970s had reignited a public debate about whether it was wise to have one corporation run two very different public services. A review group, the Carter

Committee, began in 1975 and, in 1977, recommended that the Post Office separate into two corporations, one for Post and one for Telecom. In 1976, the department thus commissioned a model of the Post Office corporation to gain a holistic understanding of the telecommunications business, as it became increasingly likely that the telecom business would detach from the Post Office.[56] The department engaged David Probert, a researcher at Cambridge University's Department of Control and Management Systems, to develop a system dynamics–based model of the telecommunications side. As a student, Probert had interned at the Post Office's research station, Dollis Hill, for the summer of 1970, working on a statistical analysis of signals and noise in digital communication systems. From 1973 to 1976, Probert completed his PhD research on stochastic machine learning at Cambridge and then transitioned directly into working for the Post Office on a system dynamics model of the telecom business.[57]

The Long Range Planning Model (LRPM), delivered in 1977, thus explored "alternative corporate futures."[58] Written in FORTRAN and run on an IBM 3033 time-sharing system, the model grouped the business into four conceptual modules: marketing, personnel, finance, and technology. Planners simulated corporate futures by altering a cluster of up to ten parameters and tracking the effects across 180 variables, showing the company's future finances, equipment needs, total workforce, and more across a thirty-year time horizon. As these outputs referred to the entire telecommunications business, not individual departments, operational departments mainly used the model to analyze the impact of strategic choices on the whole business. In this sense, system dynamics' first use, simulating holistic corporate futures, united the telecommunications business into an independent whole, separate from the Post Office.

The LRPM took a new, pluralistic approach to futurity. Probert described how the model did not predict a singular corporate future, but generated a range of alternative futures for management to "expand our own 'mental models.'"[59] The model simulated different futures: the "uncontrollable future," a predetermined future for which the model would identify the resources needed to execute short-term plans; the "designed future," which simulated "considerable freedom in controlling the corporate destiny"; the "self-fulfilling future," which assumed that a designed future would achieve "full implementation"; and finally, the "future as a game," which blended the above types to convey to management that the future would result from conflict among various corporations, each attempting to effect their designed futures and each influenced by the inertia of history.[60] This pluralistic approach shows how simulation began to change the telecom business's approach to futurity. Simulation put possible futures in dialogue with the present to aid decision-making. Long-range planning had thus changed from envisioning one digital future, as it had been under Merriman, Harris, and Whyte, to anticipating alternative futures. But in all cases, anticipation was also normative, conjuring futures of an independent telecom business.

The model's delivery in 1977 was timely, as the Carter Committee review also finished that year, recommending that the government separate post and telecommunications so that the telecommunications business could reinvest its profits, rather than support the loss-making postal business.[61] This recommendation didn't just vindicate the senior management's desire for the split but also reinforced a shift in thinking about telecom from a public service to a commercial enterprise. This enabled an understanding of the telecom business as something that could either reinvest profits or support a loss-making postal business. To the Post Office board's ire, however, the Labour government instead delayed separation for an ill-fated experiment in industrial democracy. This experiment increased the Post Office board from seven management members to nineteen, adding seven trade union members and five external members from industry.[62] The former postmaster general, Tony Benn, had advocated industrial democracy since taking over the Department of Trade and Industry in 1974. By 1977, James Callaghan had replaced Harold Wilson as prime minister, and Benn successfully secured Callaghan's approval for the industrial democracy experiment. The experiment started that year, and so in 1978, the Callaghan government vetoed the Post Office split to continue the industrial democracy experiment.[63]

The opinions on industrial democracy were mixed. Initially, the press reported it as a pioneering experiment, but soon suggested that Post Office management was undermining it, tired of industrial relations dominating the board's time.[64] Britain's interest in industrial democracy in the late 1970s, which the government only ever notably implemented in the Post Office and British Steel, has been cast as a flirtation that went as far as it did only because of underlying economic issues and the failure to find lasting solutions to those issues.[65] An academic report on the experiment by the University of Warwick's Industrial Relations Research Unit found that difficulties had stemmed from the sheer size of the Post Office organization and had been exacerbated by the conflicting expectations and interests of management and union members:

> Management members in particular claimed to take into account the interests of the various groups involved in and with the Corporation. The best means of doing so was often seen by managers to be to ensure that those interests had no direct representation at key management decision-making points.[66]

The management board members discounted the Warwick report as overly academic and "essentially anthropological" but agreed that they had lacked united purpose during the experiment.[67] Industrial democracy thus seems another episode in Post Office history that showcases the friction between the government wielding the Post Office as an instrument of economic and industrial policy and Post Office management's ambitions for greater independence from state control.

The Conservatives' election in May 1979 not only meant the end of industrial democracy but also foreshadowed telecom liberalization, which had further effects on modeling in long-range planning. James Prior, Margaret Thatcher's employment secretary, had first been reluctant to end industrial democracy, wary of upsetting the Trades Union Congress. In 1979, however, Keith Joseph, the secretary of state for industry, ended industrial democracy after receiving resignation threats from the Post Office chairman, William Barlow, who had been appointed in 1977 on the presumption that he would oversee the split and run the new telecom corporation.[68] Furthermore, Joseph also announced that the postal and telecom businesses would split and noted that this split would be part of a broader review of the telecom monopoly, setting the stage for its future liberalization.[69]

This political change highlighted a key deficiency in long-range planning's first model, the LRPM, which had modeled only the corporation and not its wider environment. Probert thus added politics and economics into long-range planning's simulations, and by 1979 he had developed a strategic control unit (SCU) for the Long Range Planning Model. The strategic control unit was a bolt-on program that enabled the Long Range Planning Model to simulate various future crises, from "economic recession" to "severe constraints on tariff increases," and the business's ability to recover from such crises.[70] Given that economic instability had also provoked long-range planners to widen their forecasting horizons earlier in the 1970s, it appears that the turbulence of industrial democracy, further fiscal constraints from the 1976 IMF bailout, and the looming threat of liberalization also influenced the SCU's development. The unit allowed users to set objectives for corporate performance parameters, and when the model initiated a crisis by "spiking" certain variables, the unit would try to normalize chosen parameters. This meant that long-range planners could map the viability of different paths from crisis to recovery, and, tellingly, Probert envisioned the SCU as guiding management decisions within real-life crises.[71] In effect, the SCU took a normative turn by turning the LRPM from a model that described various futures into a model that prescribed specific futures. This idea that predictive computing could prescribe the future would transform telecom managers' visions for a digital network as they confronted liberalization.

FROM 1980 TO *NINETEEN EIGHTY-FOUR*

Computing and prediction helped BT's senior management negotiate the transformation of their visions for digital infrastructure through liberalization and privatization, which first concretely appeared on the policy agenda in 1979. After Keith Joseph announced in 1979 that the new Conservative government would review the Post Office's telecom monopoly, he commissioned a report into telecom liberalization by Michael Beesley, professor of economics at the London Graduate School of Business Studies. Beesley had previously served as chief economic adviser to the Ministry of Transport in the 1960s and had undertaken cost-benefit analyses of the M1 motorway and the London Underground's Victoria Line.[72] In the 1980s and 1990s, Beesley, along with the economist Stephen Littlechild, came to the fore as the two published together and separately on market reform, liberalization, and privatization, becoming two of the most prominent economists advising the Thatcher governments on liberalization and privatization policies.[73] Beesley's report on telecommunications first targeted value-added network services (VANS), services in which a third party would lease lines from the telecom business to provide non-voice services, such as data transmission, between clients. Beesley then expanded the report to look at leasing lines to resale voice telephony to customers, creating further competition. Finally, Beesley also looked at expanding competition into transmission and switching networks, effectively liberalizing telecom infrastructure completely so that competitors could build alternative networks.

Beesley's conclusions were unabashedly pro-competition and proved too much, too soon, for the Thatcher government. Beesley concluded that there should be no restrictions on VANS provisions, that BT should lease circuits to competitors, and that competitors should build alternative telecom infrastructures.[74] These suggestions brought strong opposition from BT, the unions, and even some business users, who preferred liberalization to focus on VANS provision and private circuits rather than competing systems for residential customers.[75] The government chose a compromise, inviting proposals for a single alternative network, which formed in 1981 as Mercury Communications, funded by a consortium of Barclays Merchant Bank, BP, and the telecommunications company Cable & Wireless. Mercury gained its initial license in 1982 to supply leased private circuits and then became a full alternative system in 1984, gaining a license as a public telecommunications operator.

Meanwhile, preparation for the split went ahead. The government renamed the telecom business British Telecom, which reorganized itself into a market-oriented structure to prepare for liberalization.[76] The board created a new directorial position, Organisation and Business Systems Development, which combined long-range planning with the Management Services Department. Showing the business's greater focus on market opportunities, BT renamed management services as business systems, and long-range planning merged with business planning to become the Business Planning and Strategy Department (BPSD).[77] Alex Reid, the head of long-range planning, became director of business systems, while Probert won a full-time position within the BPSD as director of strategic modeling.

With this new market focus came a new market-oriented model. After the government announced that the telecom monopoly would end, Probert developed the Integrated Communications Demand Model (ICDM) to simulate BT's competition and market share.[78] This model's simulations tracked 150 indicators across a thirty-year time horizon, but differed from the Long Range Planning Model in that it grouped indicators into three areas of market demand—terminals, connections, and services—varyingly allocated to BT or its competitors throughout the simulation. When Probert completed the model in 1981, he emphasized how this model showed long-range planning's commitment to this new marketized future by declaring that "competition and diversity are ideas of the future."[79] Probert used the ICDM to school BT staff in these so-called ideas of the future, describing that demonstrating the model to a manager "can be a valuable stimulus to more flexible thinking on the questions of market demand."[80] Probert wrote articles about the model for company magazines and journals, produced brochures, and arranged presentations, seminars, and drop-in meetings, using the model to educate managers about the market. He also emphasized the model's color interface, a new addition, as instrumental for its pedagogical usage, explaining that the new interface would allow the presentation of simulations "in a neat and compact manner which is acceptable to management."[81] Probert described how he could use colored curves, bar charts, and numerical values to facilitate a management-friendly output of the model's analyses, concluding that "the extent to which managers are prepared to entertain model-based approaches is significantly affected by the 'friendliness' of the interface."[82]

The reorientation of simulation from decision-making input to training tool responded to staff resistance to liberalization and privatization. The board favored liberalization, as it would mean greater freedom from state economic controls, and this continued with the board's support of privatization, which freed BT from public-sector borrowing restrictions. Many staff, however, were not so compliant. The Post Office Engineering Union began anti-privatization action on several fronts: working-to-rule on international traffic; refusing to interconnect to BT's new competitor, Mercury; and refusing to maintain equipment used by Mercury and its owners, Cable & Wireless, BP, and Barclays.[83] The board held a crisis meeting in April 1983 with the POEU, in which union leaders emphasized their grievances that the government had unexpectedly liberalized international telecommunications alongside domestic, allowing Mercury to enter international telephony. While this decision also vexed BT's board, which had assumed that BT would remain the sole carrier of international traffic from the UK, the board could not persuade the POEU to cease action.[84] Industrial action thus continued, and, by October 1983, BT had suspended two thousand engineers.

The Integrated Communications Demand Model formed part of several tactics, supported by computerized systems, that BT's senior management used to convince staff that liberalization and privatization were positive changes.[85] These tactics included letters posted to staff's homes and discounts on employee shareholding. For example, BT gave access to its first computerized central employee database, PRISM, to Hill Samuel Registrars, BT's employee share scheme administrator, to build a register for direct mailing about employee share ownership.[86] This strategy even extended to holding senior staff meetings with Patrick Jenkin, Keith Joseph's successor as minister for trade and industry, who explained to staff the necessity of privatizing BT. The board circulated Jenkin's message to all managers to communicate these explanations to their staff and "calm any exaggerated fears."[87] Communiques in this style continued in the run-up to privatization, with staff informed that competition would positively stimulate BT and reminded that, with the Conservative victory in the 1983 general election, any industrial action against privatization would be seen as defying the will of the electorate.[88]

Computer simulation's shift from planning to pedagogy reveals its importance to British Telecom. Organizational simulations become truly productive once they turn from representations into managerial tools.[89] This change happened with system dynamics in BT, which Probert developed from the Long Range Planning Model, a corporate representation, to the Integrated Communications Demand Model, a training tool, as part of a broader strategy by British Telecom's management to marketize staff. Probert often emphasized his "marketing" of the ICDM, from its drop-in clinics to its colorful, manager-friendly interface, but this marketing extended beyond the simplistic sense of advertising its use to staff.[90] System dynamics in British Telecom was also "marketed" through the progressive incorporation of free-market principles into its code and "marketized" staff by teaching free-market views. The ICDM's usage also further highlights how long-range planners' attitude to the future had changed. This model extended the SCU's prescriptive approach, overtly displacing the norms of monopoly with the future of a competitive, marketized network. The ICDM taught managers to displace the norm of public ownership with norms of free markets and competition. This predictive, pedagogical power was reinforced by the notion that these computer models "scanned" the future.[91] A powerful symbol of this predictive authority appeared in the logo given to BT's new Business Planning and Strategy Department, an all-seeing eye that gazed into the future (figure 2.1).[92] This inverted Harris's view of long-range planning from the department's founding, where the goal was explicitly to "invent the future, not to predict it." Where the department's early activities and models served to uphold the Post Office's vision of a universal digital infrastructure for Britain, by 1981, BT's prediction technologies instead surveilled the future for both threats and opportunities posed by the market.

Long-range planning at BT also forged this new relationship between computing and futurity through computers' symbolic role in new visions for Britain's telecom infrastructure. In November 1980, the BPSD organized

FIGURE 2.1
The Business Planning and Strategy Department's new logo. Source: TCC 75/2, BT
Archives. Courtesy of BT Group Archives.

a weekend retreat, "Into the 21st Century," for senior management and
board members to prepare for liberalization.[93] This seminar, taking place
over two days and attended by twenty-eight figures from BT's board and
management, became a venue for BT's most senior managers to articulate
new visions of societal, organizational and technological change, visions
that stood in stark contrast to those articulated by Merriman and Harris
thirteen years earlier. Indeed, Peter Benton, BT's managing director, introduced the seminar as a challenge to both embrace the future and discard
the past, explaining that it was an "attempt to appraise the forces which
will affect our destiny; and perhaps identify the forces which we ourselves
will need to exert to ensure a prosperous future" and that to do so, it was

necessary "to ensure that we were not blinkered by our traditions and preconceptions."[94]

Much discussion at the retreat concerned the power of computing to transform the economy and the state. Alex Reid, the director of business systems, long-range planning's parent division, spoke of how networked computing would give successful enterprise "a vision of the total scope of the market" and the "freedom to perform any function in any country."[95] A group discussion explored the potential for portable computing to facilitate a diverse and competitive market and render the labor force more mobile.[96] In another echo of Bell's postindustrial society, Richard Greensmith, head of the Telecommunications Industrial Relations and Safety Department, spoke about how the "electronic office" and other computerized systems would "automate sole [sic] destroying jobs."[97] The discussants extended this vision further to suggest that intelligent machines could abolish the need for human labor altogether, asking, "Couldn't an intelligent machine program itself well enough?"[98] Another discussion concerned the future place of information as "the basic commodity of service industries" and the difficulties in evaluating the "output" of information compared to the solid material output of "iron bars or bushels of wheat." The discussion concluded that in monitoring and allocating the information commodity, "governments should . . . take the lead; or market forces should be allowed to work unfettered; but it is illogical to argue for both simultaneously."[99] Here, BT's senior management saw postindustrial computerization as presenting Britain with a binary choice between government control or liberalized markets and private ownership. Privatization and liberalization seemed, to BT's senior management, two sides of the same coin, and the idea of competing as a publicly owned enterprise was never entertained.

These views became even clearer as they further related digitalization to individualism and small government. J. J. Wheatley, BT's head economics adviser, imagined how computerization and telecommunications could liberate the state from big government. He suggested that microtechnology, which would generate high-value manufacturing and service-based industries, would revive the economy and combat inefficient, swelling bureaucracies:

> There could be a convergence of computing and communication technology, with "small government" aspirations:
>
> - Small is beautiful.
> - Small is cheaper.

- Large is unnecessary.
- Devolution gets government closer to people.
- Small is anti-bureaucratic.[100]

Wheatley's sentiments latched onto a broader vogue for technology and decentralization inspired by works such as the economist E. F. Schumacher's 1973 book *Small Is Beautiful*, although Schumacher's book focused more on "appropriate technology" and sustainable development.[101] Wheatley's focus on small, cheap government, however, was more typical of the British state's contraction under Thatcher and echoed digital libertarian ideologies that saw digitalization as an opportunity for small government and a flat, decentralized society.[102]

Normative futures of emancipatory individualism joined these political-economic aspirations. Richard Greensmith, for example, continued by explaining that the electronic office and personal networked computing would mean "a trend away from bigness and centralisation, . . . thus restoring the importance of the individual."[103] Charles May, BT's director of research, envisioned that mobile telephony would overcome "the tyranny of the local line":

> I am convinced that the next generation of businessmen—or perhaps the next but one—is going to want a truly universal pocket telephone. . . . This would completely release him from this "tyranny of the local line" and enable him to make and receive calls wherever he happened to be.[104]

This collision of information technology and individualism was part of a broader trend, both in Britain and abroad. Kenneth Baker, Thatcher's minister for information technology, argued that computerization under the Conservatives would provide greater personal freedom, contrasting this with computerization in the totalitarian hands of the "Electronic State."[105] On the left, Tony Benn suggested that, in the hands of the worker, information technology could give "people a sense of freedom," although he warned that the microchip, in the hands of the capitalist corporation, was "tyranny in the form of liberation."[106] These readings of digital technology again echo the digital ideologies that took hold among US politicians and technologists from the 1970s, who similarly interpreted personal, networked computing as facilitating personal freedom.[107]

These visions, however, focused not only on liberation but also on surveillance. BT's senior managers and engineers saw computers as powerful predictive machines that surveilled not just the future, but also BT's users. Charles May explained that global computer tracking of users could enhance mobile telephony's emancipatory power:

> I believe there are about 6 thousand million people in the world. . . . I see no problem of keeping track of them all so that the international telephone system— already the most elaborate and complex thing man has ever created—can find and call anyone in the world wherever he may be.[108]

Charles May, BT CEO George Macfarlane, and Roy Harris also discussed the potential for a heuristic machine-controlled network that could shape users and predict their demands and desires:

> Although it is argued that technology is only justified if it serves people and that it should not be self-perpetuating, the increasing power and complexity of machines shapes people's demand and desires. Human desire for information is not random: the assembly and accessing of databases reflects users' interests and heuristic machines will judge what individual users are most likely to want to know.[109]

The discussion even envisioned that these predictive surveillance machines could gift BT the power to transform society itself. Participants concluded that using computers for "the selection and manipulation of that information" meant "there is potential for moulding society by selecting the contents of the databases."[110] By 1980, computers had not only changed long-range planning from inventing to predicting the future, but also appeared close to realizing the "bleak mechanistic prospect" that J. S. Whyte had warned about eleven years earlier.

BT's senior management recognized these visions' darker undertones. Replacing human labor with intelligent machines was a "heretical idea," while using predictive computer databases to mold society had "ethical problems."[111] Charles May argued that his global computer tracking system would be permissible so long as users could opt out, a "god-given right" that he, "as a technologist, would defend to the death," and in doing so, he explicitly distanced himself from a "big brother" approach.[112] Distancing their visions from science-fiction dystopias appears to have been a popular strategy among BT management. J. J. Wheatley also labeled dystopic interpretations of these computerized futures as the "politics of the pessimists," suggesting that the "dictatorship of technology" of H. G. Wells's *The Shape of Things to Come* and "the enslavement by the information society" of George Orwell's *Nineteen Eighty-Four* were lazy cultural references that bore no resemblance to BT's visions.[113]

The ways BT managers contrasted these science-fiction futures to their new digital vision of predictive computer control reveals much about the politics of this new vision. Wheatley invoked *The Shape of Things to Come* as a dictatorship of technology, but *The Shape of Things to Come* follows a Wellsian convention wherein societal collapse is a necessary precondition for the emergence of utopia, and this utopia emerges through the guidance of scientists and technocrats that command powerful technology.[114] In *The Shape of Things to Come*, these are the technocratic airmen who control aircraft. Referencing this book perhaps suggests that Wheatley saw this vision of small government and corporate technocracy, enabled by digital convergence and predictive computing, as acceptable so long as engineers and managers remained in control. May and Wheatley distanced BT's visions from Orwell's "big brother" future and the "politics of the pessimists."[115] The contrast is easier to see here. *Nineteen Eighty-Four* is often invoked to describe contemporary, high-tech, corporate surveillance, but Oceania is a violent society with a centralized state and quite visible, low-tech modes of surveillance, such as cameras and telescreens.[116] BT's management, however, read *Nineteen Eighty-Four*'s dystopia as irrelevant to the competitive marketplace and the private sector. In contrast, they proposed invisible, computerized modes of corporate surveillance that paradoxically would support their new vision of digitalized decentralization, individual freedom, and small government.

This paradox shows the political power of prediction technologies and the "hypersurveillant imaginary" that they offer. In the postwar era, futurology's prediction technologies, which included techniques like system dynamics, became the "core political technology of the present," a tool of technocracy that allowed direct intervention in the future.[117] By turning the long-range into a category of rational decision, long-range planners could manage plural futures, simultaneously affirming the future as a realm of liberal choice while also extending technocratic control into the future. Similar values are contained within the hypersurveillant imaginary, a particularly popular sociotechnical imaginary under late capitalism.[118] Whereas traditional surveillance monitors the present, hypersurveillance uses the predictive power of computing to put the future under surveillance, so that present action can preempt future change.

At the 1980 long-range strategy seminar, the discussions about using heuristic, predictive computing to anticipate and shape customers' data consumption matches the predictive corporate, consumer surveillance that hypersurveillance describes. The appeal of both hypersurveillance and prediction technologies emerges from the political and commercial power they promise to their wielders, claiming that one can intervene in the present by picking futures. The history of long-range planning at BT is largely a history of political prediction technologies, from early models of exchange obsolescence and trunk digitalization to more sophisticated corporate system dynamics models. The 1980 long-range planning seminar, however, shows the emergence of the hypersurveillant imaginary, in which BT's managers, engineers, and long-range planners discussed the power that hypersurveillance would give them over consumer markets. By 1980, long-range planning at BT had collided prediction technologies with the hypersurveillant imaginary. Strategic modeling allowed BT to choose a future of "competition and diversity," in which digital convergence would facilitate small government, free markets, and individual freedom, and within that future, the hypersurveillant imaginary offered a way to continue to "mould society." In this new digital vision, BT could avoid concerns about privacy and totalitarianism altogether by using prediction technologies to rewrite the future, averting the path from 1980 to *Nineteen Eighty-Four*.

CONCLUSION

The history of long-range planning at the Post Office and BT shows how futurology served three roles. First, long-range planning sustained Merriman and Harris's vision of a universal digital network. After the failure of Britain's planning moment during the 1960s, attention turned to new ways that the state could know the future. Under the advice of McKinsey, the Post Office founded its Long Range Systems Planning Unit, and engineers such as Harris and Whyte made plans that reaffirmed the Post Office's commitment to an integrated digital network and showed their sensitivity, as operators of a nationalized infrastructure, to these new technologies' privacy risks. Underscoring the department's importance, many of its senior figures earned promotions after working on long-range planning. Roy Harris went on to head the Telecommunications System Strategy Department, responsible for the flagship switching and transmission projects (covered in chapters 3 and 4). J. S. Whyte was promoted to head the Operational Programming Department, where he negotiated a national controversy about Post Office computer modeling (explored in chapter 3) and, after that, succeeded Merriman as the senior director for engineering. Alex Reid became director of business systems and (discussed in chapter 7) helped orient BT's digital network toward financial users from the City of London.

The department was an incubator not only for engineers, however, but also for technologies. Reflecting a national political economy that picked industrial and technological "winners," Post Office long-range planners too picked Viewphone as a technological winner, imitating AT&T's Picturephone. But Viewphone also emulated AT&T's Picturephone by successfully failing, never reaching the market, and yet standing as an icon of the Post Office's goal for an integrated digital network that transmitted voice, data, and video. Long-range planners also began to use computers models to pick winners, as the ALEM model "chose" TXE4 for the Post Office's next main telephone exchange, while the UKTTF model "chose" the waveguide as a high-bandwidth transmission technology for Britain's information highways. Chapters 3 and 4 follow, in part, TXE4 and the waveguide, showing these choices' consequences for the history of digital switching and transmission.

Futurology's second role was in giving an entry point for new computing ideas and techniques. As long-range planning became more expansive during the economic and energy crises of the early 1970s, planners studied new ideas about how computing and communications would transform society. While practical in their aims, undertaking surveys and writing reports on telecommuting and working from home, these studies also looked to the ideas of academics and futurologists on the relationship between technology and society. Foremost among these was Daniel Bell and his "post-industrial society," which reinforced planners' expectations of an information revolution and reaffirmed their use of computers as prediction technologies. In some ways, this was as an extension of Post Office engineers' first encounters with information theory and cybernetics, when they became enamored with the idea of a "second industrial revolution," wherein "human beings, used as sources of judgement, may also be replaced by machines."[119] Indeed, Bell himself pointed to cybernetics as a component of "the computer age," although he also criticized cybernetics as a mechanistic and closed vision of society.[120] Regardless, Post Office engineers and long-range planners paid attention to scholarly expectations about digitalization and used these expectations to bolster their plans for a universal digital network. During this period, the department looked to more sophisticated computer models for long-range planning, also influenced by computer simulations of energy and resource scarcity such as *Limits to Growth*. System dynamics played a central role in long-range planning, first in helping cohere telecom into a corporate whole, separate from the postal service and, second, as a pedagogical tool used to inculcate free-market values in staff during liberalization. At first, long-range computing reinforced a singular corporate future of digital infrastructure monopoly, and then it embedded free-market values for a liberalized future.

Futurology's third role was as a chrysalis within which the Post Office's universal digital vision could transform from a vision of monopoly into a vision of markets. At the 1980 long-range planning seminar, BT planners, engineers, and executives anticipated a future in which predictive computers, embedded throughout Britain's telecom infrastructure, would give them a complete vision of the market and their customers. Allowing BT to monitor, predict, and filter customers' data usage, predictive computing would give BT the power to mold society itself. This new vision was so powerful because it offered a way to use digitalization to negotiate a new marketplace and track and tailor customer's individual needs in a world seemingly defined by individual choice and small government. Long-range planning was not just a venue through which managers and engineers articulated this new vision. Prediction technologies showed new ways to think about futurity that permitted managers and engineers to articulate these contradictory visions. How could BT's senior management simultaneously prize choice, competition, and small government while envisioning a future of predictive computer control and surveillance? Prediction technologies, pioneered within the telecom infrastructure by long-range planners, showed how computer control and surveillance of both customers and the future could preserve and optimize choice.

This chapter already begins to indicate the ways in which digitalization intersected with changing political economy. Long-range planning gave a venue for managers and engineers to develop and deploy digital tools and imaginaries that engaged with wider political economy. In the 1970s, engineers and planners imagined a nationalized digital network for information services, while acknowledging the risks of state-owned digital networks. They also used digital tools, in the form of computer simulations, that reinforced the Post Office's national monopsony over equipment procurement. In 1980, these engineers and managers instead imagined the commercial opportunities of predictive, commercialized computer control and surveillance in a privatized infrastructure, while their computer simulations turned away from monopoly and monopsony and to competition and liberalization. Crucially, these visions and tools were not just reflections of the political economic environment. They co-constructed these changes to the political economy of telecommunications. Models reinforced monopoly, corporatization, and liberalization. Digital imaginaries about commercial surveillance helped senior management understand the opportunities that digitalization could offer them in a marketplace. Digitalization in long-range planning did not merely follow political economy but shaped it. This does not mean, however, that this is a history of ruptures alone. There was continuity alongside change. The Post Office's original digital vision placed computer control at the center, both as managerial tools and as control centers within Britain's future digital infrastructure. So, while BT executives' 1980 digital visions differed from the past and reflected the changes of the early 1980s, they also maintained the core value of computer control that was at the heart of Merriman's digital vision from the 1960s. The next chapter investigates how the Post Office and BT digitalized Britain's telephone exchanges, embedding this computer control into Britain's telecom infrastructure.

II PROJECTS

II   PROJECTS

3 MONOPSONY IN THE MACHINE: SUPPLIERS, LABOR, AND TELEPHONE EXCHANGES

As the last two chapters have shown, a self-governing network, realized through computer control, was fundamental to Post Office engineers' vision of a national digital network and to BT managers' understanding of their digital, liberalized future. But this vision would require computerizing the switching equipment in telephone exchanges, which route traffic from senders to receivers. This move to electronic and digital switching equipment from the 1950s to the 1980s shaped and was shaped by public ownership and monopoly. A history of this move is especially important for understanding how digitalization influenced the change from a nationalist to neoliberal political economy because, in this case, the Post Office was not just a monopoly. It was also a monopsony, being the sole purchaser of the switching equipment and the labor, both technicians and telephone operators, that made these exchanges work. The question of how the Post Office and BT computerized these exchanges is thus inseparable from its status as a monopsony.

The supply of public switching equipment is key to the history of the telecom industry. This equipment was the "largest single item" in the telecom network and the "flagship product" of national telecom industries.[1] At first, switching was done manually, by telephone operators, but moving to automated switching equipment offered opportunities to rewrite the dynamics of national telecom markets. For example, during North America's Gilded Age, local, independent telephone companies automated switching quickly to oppose the hegemony of the Bell System, which automated slowly, preferring its telephone operators rather than customers to control traffic.[2] In the UK, automatic mechanical switches were introduced in 1912, after the

Post Office took over the National Telephone Company and telecommunications entered public ownership.[3] For most of the twentieth century, the Post Office procured automatic switching equipment via "bulk supply agreements," which were effectively a cartelized relationship between the Post Office and its main suppliers to provide standardized equipment and share R&D resources. This was a traditional model in Western Europe, known as the "telecom club," in which the national PTT and a small group of domestic "court suppliers," closed to foreign entrants, ran the nation's telecom manufacturing industry.[4] This hallmark of a nationalist protectionist political economy has been called the "postal-industrial complex."[5] But during the period covered in this history, that dynamic began to change. Rising user demand, combined with insufficient equipment supply through the 1960s, meant that the Industrial Reorganisation Corporation, created by Harold Wilson's Labour government, ended the bulk supply agreements, theoretically opening the equipment supply market to new entrants.[6] This chapter thus looks at how the Post Office's monopsony over equipment supply intersected with digitalization through the liberalization and internationalization of public switching markets.

The Post Office had a monopsony not only over the market for the equipment inside telephone exchanges, but also, effectively, over the market for the labor in those exchanges too. As the sole national telecom operator, telephone operators and exchange maintenance technicians had little choice over their employer, leading to a power dynamic among the Post Office, these labor forces, and automation that gives further insight into how digitalization influenced the end of not only the Post Office's monopoly but also its monopsony. Labor history has long been connected with history of technology, particularly given the role of technology in automating the factory floor, the quintessential site of labor history.[7] But that is not to say that the role of labor in communication networks has gone overlooked.[8] Messengers, operators, line workers, and maintenance technicians are the "internetworkers" that make these communications networks work.[9] In national telecom networks, telegraph "messenger boys" and telephone operators have received the most historical attention, which has shown how these labor forces actively produced these systems and were then automated away.[10] Exchange maintenance technicians, on the other hand, have received less attention, although the recent turn to a history of maintenance is addressing this.[11] The Post Office also attended to the operator and technician labor forces differently, not least because they were unionized differently. The Post Office Engineering Union, a technocratic union with origins in white-collar trade unionism, represented the technicians and engineers, while the Union of Postal Workers, with roots in guild socialism, represented the telephone operators, who were by far the largest labor force from the telecom business within the UPW.[12]

But unions were not the only difference between these two groups. This labor history is also a gender history, as the technicians were nearly all men, while the operators were mostly women. Gendered labor has been central to the history of digitalization. The vital role of women in developing computer hardware and software was overwritten by "systems men" and "backroom boys," who particularly undermined the British computing industry as it lost a highly skilled labor force of women "computers."[13] Telecommunications history tells a similar story of pioneering women, who were then erased and misrepresented.[14] Women were keen and early telephone users, although early telephone advertising all too often deployed gendered stereotypes of domestic life.[15] Women were influential not just as telephone users but also as telephone operators.[16] In both the US and the UK, national monopolies played a decisive role in delineating men and women's roles in telephone operation, as AT&T and the Post Office feminized and racialized operator labor, idealizing the telephone operator as a white, virtuous woman, pious and modest.[17] The gendered history of these work forces matters to the history of the Post Office's monopsony over its exchange labor.

Understanding how the Post Office implemented computer control in its switching centers, which is inseparable from this monopsony, thus means examining the history of both supply and labor through three case studies. The first case study examines GRACE, the "robot telephone operator" and one of the Post Office's first major postwar automation projects, a machine designed to automate long-distance calling in the 1950s, and focusses on how gendered labor hierarchies structured this early automation. The second case study turns from labor to supply, exploring the Post Office's purchase of a new series of electronic exchanges, TXE4, from one of its suppliers, STC, after the IRC ended the bulk supply agreements and liberalized equipment supply. This case study continues the history of ALEM, A Local Exchange Model, commissioned by J. S. Whyte in the Long Range Planning Department to plan the rollout of new electronic exchanges, and shows how the computerization of this decision-making process triggered a national controversy among the government, the Post Office, and its suppliers. This case thus shows how the Post Office's vision of computer control was not confined to the interior of the exchange and extended into exchange planning too. The third case study investigates System X, Britain's first digital, computerized telephone exchange, from the perspectives of both supply and labor. It looks at how supplier issues plagued System X's development, culminating with government intervention during Thatcher's first term, and how the computerization of telephone exchanges transformed the labor of both technicians and operators, paving the way for a massive program of redundancies during the 1980s and 1990s as BT attempted to become "leaner" for global markets.

GRACE, THE ROBOT TELEPHONE OPERATOR

On December 1, 1958, at a televised ceremony in Bristol, Queen Elizabeth II placed Britain's first automatic long-distance call, calling the Lord Provost of Edinburgh. Subscriber trunk dialing, known by the unfortunate acronym STD, was the Post Office's name for this new service, whereby a telephone user could place a long-distance "trunk" call without an operator's assistance, as machines would route the call automatically. In newspaper reports, Movietone and Pathé newsreels, and its own publications, the Post Office proclaimed GRACE, the "robot telephone operator" and "electronic brain" that made STD possible.[18] GRACE was not a robot, but a register-translater, which registered dialed numbers and translated them into a form recognizable by the switches in a telephone exchange. GRACE was at the forefront of electronic developments in British telecommunications. While Highgate Wood, explored in chapter 1, was a prototype all-electronic exchange that never reached full-scale production, GRACE was an electronic component implemented in electromechanical telephone exchanges across the country to enable automatic long-distance calls. The history of GRACE, the machine with a woman's name, reveals how automation, gender, and monopsony met in Britain's telecom infrastructure in the late 1950s.

Investing in Britain's long-distance telecom infrastructure was a national priority for the Conservative governments of the 1950s. World War II had shown the strategic importance of trunk communications between cities, but during the war, the Treasury had taken over financial control of the Post Office and, after the war was over, refused to give it up. A parliamentary committee backed the Treasury in 1950, arguing that Britain's strained economic condition meant that it was impossible to return to the prewar arrangement, in which the Post Office controlled its finances but made an annual £10.75 million payment to the Treasury for the public purse.[19] While Post Office finances remained under Treasury control, the first postwar telecom bill, the 1952 Post Office and Telegraph Money Bill, prioritized trunk spending to ensure resilient connections between cities.[20] This was not enough, however, to stave off poor commercial performance or to invest in automating long-distance dialing. By 1955, automation's financial needs prompted a new relationship between the Post Office and the Treasury. The government instituted a five-year trial of the prewar model, in which the Post Office would manage its finances independently and pay £5 million per year to the Treasury, reinvesting any excess profits.[21]

Automation was a matter not just of policy but also of publicity. Crazes and scares about automation defined the 1950s. New ideas about an age of automation and electronics came to the fore, manifesting in the widespread popularity of robots and "electronic brains."[22] This popularity extended into the 1960s, as Harold Wilson's 1963 "white heat" speech at the Labour Party's annual conference promised to harness automation for the national good, while Leon Bagrit, chairman of Elliott-Automation, delivered the BBC's 1964 Reith Lectures on the "age of automation."[23] At the same time, there were widespread fears about the impact of automation on jobs. In 1956, for example, motor workers at Standard Motors went on strike against automation, which the *Daily Herald* called the "men against robots" strike. Alongside the robots craze, automation's public role in the late 1950s was thus also beset by fears about redundancy, worker resistance, and "visions of jobs disappearing into factories without humans."[24]

This perhaps explains why the Post Office humanized its robots by naming them and, in the process, gendering them. GRACE was not the Post Office's first robot. In 1936, the Post Office had launched TIM, the speaking clock. TIM was voiced by Ethel Cain, also known as "Jane Cain," a connection clerk and former telephone operator, who won the "golden voice" competition to select TIM's voice. John Masefield, the poet laureate and competition judge, described Cain as having "one of the most beautiful voices I have ever heard," while the *Post Office Magazine* described Cain as "a blue-eyed slim blond in the mid-twenties."[25] Cain's feminine appearance and voice were apparently important qualities for telephone customers and magazine readers, yet TIM, the speaking clock, was given a male name despite using female labor. The Post Office deployed another male robot, ERNIE, the year before GRACE debuted. ERNIE, launched in June 1957 and designed by Tommy Flowers, the engineer behind Colossus and Highgate Wood, was a computer that randomly chose prize winners among premium bondholders. ERNIE's most recognizable feature was its operation console, designed with switches, lights, and oscilloscopes "to foster the public impression of a 'modern scientific robot.'"[26] ERNIE, however, was quite inscrutable to the public. Only its console was visible, and the numbers it calculated were not immediately disclosed but passed to a back room for sorting, showing how ERNIE and its clerical staff enjoyed considerable public trust.[27] Furthermore, ERNIE's name was a deliberately formed backronym (for "electronic random number indicator equipment"), named after the postmaster general, Ernest Marples. ERNIE thus stood in for not only the trusted authority of government clerical staff, but also the trusted authority of the male politician.

Two more gendered robots, ELSIE and ALF, also debuted around the same time as GRACE and ERNIE. ELSIE, the "electronic letter sorting and indicator equipment," sorted letters, while ALF, the "automatic letter facer," oriented letters during postal sorting. A 1957 Post Office magazine contributor revealed the extent that gender played in presenting these robots: "All of us have heard of ERNIE and his girl friend GRACE. Now, even I know what ERNIE stands for, and what GRACE stands for is surely none of our business."[28] The contributor continued his analogy of leading male robot with a subservient, feminized partner with ELSIE and ALF: "ALF (bless his loving heart!) spends all his time sending stacks of letters to ELSIE. . . . Not all of the letters are love letters but ELSIE treats each one with loving care." The article concluded with a sexist joke about ALF's electronic eye being a "roving eye," but that "nice girls don't mention such things." Reinforcing heteronormative relationship stereotypes, the Post Office "partnered" ELSIE with ALF and ERNIE with GRACE to publicize postal automation and telecom research, respectively. The ways that the Post Office gendered its robots shows that automation in the 1950s, which the Post Office claimed to pursue as part of the "radical advances" of an "electronic age," only served to stymie social progress.

GRACE did more than just expose gendered representations of labor and authority. It also forced telephone operators to redefine their value as laborers. GRACE's launch removed operators from long-distance dialing, and the Post Office acknowledged that this was the start of a process that would reduce operator numbers by half by 1970.[29] This prompted a Post Office report on customer service, *Telephone Service and the Customer*.[30] Ernest Marples, the postmaster general, announced this report in the House of Commons by declaring that "In this age of mechanisation, we must never forget the importance of human personal service."[31] Based on a report on customer relations from a Post Office management visit to AT&T's long-distance telephone service in New York, the report set new goals for the telephone business that linked automation, monopoly, and gender. These goals emphasized that as a public monopoly free from "the challenge of competition," the Post Office's duty was to "please the customer" with "the finest and most courteous service in the world." The Post Office would compensate the public for their reduced personal contact by asking remaining operators to behave in a more "friendly" manner. Nan Whitelaw, the UPW's assistant secretary for telephones, who had gone on the New York trip and helped plan the "friendly campaign," and who had been a switchboard operator herself, explained that this new "courteous service" involved allowing operators to emulate the casual yet friendly and courteous habits of US telephone operators. For example, operators could now share their first names with customers, and could use more informal vocabulary. Yet, as Whitelaw explained, there was little choice for operators in this new attitude: "The friendly telephone service is your only salvation in a mechanised age. We must keep our customers, because without them we have no jobs."[32] This was a labor force that, as the telephone business's main customer-facing staff, already had to conform to national, racial, and class-based standards of femininity.[33] Yet, as the "friendly telephone" campaign shows, they still had to reinvent themselves in new, gendered ways to justify their labor in the face of automation. GRACE was not only the name for a machine, but also a reminder of the virtues these women were forced to live and work by. The Post Office's monopsony gave them no alternative.

TXE4 AND THE GIANT COMPUTER MODEL

Just as new electronics had given the Post Office an opportunity to redefine its monopsony over telephone operator labor with GRACE, so the Post Office used microelectronics and computerization in telephone exchange supply to redefine its monopsony over its suppliers. This had roots in the plans that the Post Office made for Britain's digital communications infrastructure. Shortly before Merriman outlined his vision in 1967 to the Institute of Electrical Engineers, Roy Harris proposed "Project ADMITS" to Merriman. ADMITS, an "Adaptable Dispersed Modular Integrated Telecommunications System that *admits* change," which Harris developed as one of his first projects at the Long Range Systems Planning Unit, would be the technical basis for the general-purpose digital network.[34] ADMITS' basic principles were computerized exchanges that used microelectronics for switching combined with integrated digital transmission, switching, and signaling.[35] The project would proceed in tandem with industry, and its main priority would be developing a modular digital telephone exchange composed of general functional subsystems. This modularity, Harris argued, would enable an evolutionary network, as the Post Office could develop and procure components independently.[36] Merriman first received the proposal in November 1967 and confessed that he was "much attracted to the imaginative plan" but was not entirely convinced by its scope for competitive procurement, suggesting that "some elements of industry will feel that some element of industrial competitive freedom is lost."[37] Harris redrafted the proposal and the Managing Director's Committee for Telecommunications—comprising Edward Fennessy, the managing director; Merriman, senior director for engineering; and the senior directors for customer service, purchasing and planning, and finance and personnel—subsequently approved Project ADMITS. The most immediate consequence was that, in 1968, the Post Office established the Advisory Group on Systems Definition with industry to develop the technical specifications for digitizing the telecom network, especially the specifications for a future computerized telephone exchange.[38]

In the meantime, however, the Post Office needed a new series of electronic exchanges that could replace Britain's dated mechanical exchanges and serve as a halfway step to a digital computer exchange. After the failure of Highgate Wood, the Post Office–industry Joint Electronic Research Committee had abandoned the analogue approach of pulse-amplitude modulation combined with time-division multiplexing. JERC continued developing another kind of exchange called REX, short for reed-relay electronic exchange. REX used small metallic "reeds" as cross-points for routing calls, rather than Highgate Wood's electronic logic circuits, and used discrete connecting paths for each call, unlike Highgate Wood's multiplexed paths.[39] REX entered service in 1966 in Leighton Buzzard, providing service for three thousand subscribers.[40] REX, renamed TXE1, became the basis for the TXE exchange series developed and rolled out through the 1960s and 1970s, which was the subject of an onerous industrial dispute between the Post Office, STC, GEC, and Plessey revolving around the use of computer modeling to plan the modernization of Britain's switching network.

Computer modeling has a long history in the private and public sector as a decision-making tool. Analogue models, which were physical or electrical devices, were used to model electrical networks from the 1930s, and, with the advent of digital computing, computer modeling and simulation became more common, especially in the oil industry.[41] In deploying both analogue and digital models of their natural and industrial environments, organizations found ways to manage information flows about these subjects and generate supposedly accurate and authoritative descriptions of the world.[42] The public sector also used models and simulations. This history has roots in World War II, when operations researchers developed mathematical models to simulate military strategies, leaving a postwar legacy in the public sector of using models and simulations to inform political decisions.[43] Rooting these political decisions in technologies and techniques of simulation means that these decisions were technopolitical, and that models could be used to produce "matters of fact" for political decision-makers.[44] These matters of fact have always been subject to controversy and contestation, especially in government. In Britain, the governmental culture of "discreet modernism" concealed computers' roles in public administration decisions.[45] Through the case of TXE4, an electronic telephone exchange, and ALEM, the model that the Post Office used to justify purchasing this telephone exchange, this section explores how computer modeling supported the Post Office's monopsony.

In 1971, the Post Office published plans to update the telecommunications network with TXE4 reed-relay telephone exchanges, jointly developed with STC after the successful 1966 REX reed-relay trial at Leighton Buzzard.[46] The plan, as discussed in the previous chapter, was created using a discounted cash flow analysis model developed in the Long Range Studies Division. The model aimed to replace all Strowger exchanges by 1990, simulating Strowger obsolescence and replacement with TXE4 exchanges over various periods. In January 1971, the Managing Director's Committee for Telecommunications had endorsed the simulations' conclusions, and in April 1971, the Post Office board followed suit, approving the publication of the plans for review, ahead of final approval by the end of 1972. Whyte began to liaise with the Operational Programming Department but was soon appointed the department's director in order to oversee the TXE4 modernization strategy. When the Post Office publicly announced the TXE4 decision in September 1971, however, this decision and the Post Office's use of computer modeling became the subject of a fierce dispute between the Post Office, on the one side, and GEC and Plessey, two of the Post Office's main suppliers, on the other.

On January 12, 1972, the Post Office held a meeting with GEC and Plessey management, who voiced their concerns about the TXE4 strategy.[47] GEC and Plessey argued that their jointly developed electronically controlled Crossbar system, known as 5005, was cheaper and more reliable than TXE4 and yet had been rejected. TXE4 had undergone only limited experimental trials and had been developed mainly using experimental computer simulation, different from the model used for the Post Office's modernization strategy.[48] Arnold Weinstock, GEC's chairman, complained that "computer simulation alone should not be a basis for major investment decisions," while Merriman and Whyte defended their simulations. Merriman argued that extensive computer modeling had supported TXE4's design and that "it was not sensible to attempt to produce replicas of the exchange at every stage of design development, and computer simulations had therefore been employed instead."[49] Whyte, meanwhile, defended his planning model, retorting that the model "cannot of itself come to conclusions or take decisions."[50]

The Ministry of Posts and Telecommunications (MPT), formed in 1969 as the Post Office's regulatory government department after the Post Office left the Civil Service, soon became involved. The Post Office informed the ministry that "it is becoming more and more common in advanced technology to move straight from computer simulation to hardware."[51] The Post Office also pointed out that Plessey's Crossbar 5005 had its flaws. Its physical design meant that maintenance access involved high wear to cables and wires, while its limited flexibility meant that, beyond a certain point, technicians could add extra lines only by changing customers' numbers. This row between the Post Office, GEC, and Plessey soon gathered national attention, featuring in the *Economist*, the *Daily Telegraph*, and the *Financial Times*, which highlighted the central role of the Post Office's "giant computer model," from which Plessey and GEC "claim to have been excluded."[52]

This model was the "main area of doubt" for both GEC and Plessey.[53] GEC questioned the model's validity and influence over decision-making, while Dr. Willets, Plessey's director of research, announced to the Treasury, which was also reviewing the Post Office's model, that he considered it a "load of junk."[54] Meanwhile, STC, TXE4's manufacturer, mobilized to support the Post Office's model, briefing its directors to respond to press interest with the statement, "The Post Office have computer capacity and expertise unrivalled either in the Industry or perhaps in any private sector of British industry. They have programmed all the factors which affect the cost to the British public of its telephone system."[55] The controversy soon reached the point that the prime minister, Edward Heath, became involved after meeting W. D. Morton, a senior GEC engineer, at an event for senior managers from the British manufacturing industry.[56] Morton complained to Heath that the Post Office's plans were harming GEC's export prospects, so Heath wrote to John Eden, the minister of posts and telecommunications, asking him to ensure that the government properly scrutinized the Post Office's proposals and its model.[57] Eden reassured Heath that he would have to approve the Post Office's proposals even after the Post Office board made its decision at the end of 1972 and that his ministry, along with the Treasury, the Department of Trade and Industry (DTI), and the Cabinet's Central Policy Review Staff (CPRS), would investigate thoroughly. A series of government reviews thus ensued, alongside an independent review commissioned by Plessey.

Whyte relied on the Treasury O&M tradition of "discreet modernism" to counter Plessey's report.[58] "Discreet modernism" describes the deliberate concealment strategies that allowed government mechanization and computerization to thrive within government. Concealing the computer meant that expert mechanizer movements, like Treasury O&M, could cast government, not computerization, as their object of action, avoiding the problematic idea of equating Civil Service work with computers. The invisible computer could thus thrive in a supposedly incidental role. The Post Office's controversial announcement that it used computer modeling to plan its modernization strategy had breached discreet modernism. So, to close this controversy, Whyte fell back on the old tradition, concealing the computer by denying Plessey access to the model. Plessey had contracted T. S. Barker, a senior research officer at the Department of Applied Economics at Cambridge, to investigate the model for its report. Without access to the model, Barker instead provided a commentary on the Post Office's report about the model. He concluded that it was "suitable, but could be substantially improved."[59] Barker highlighted the model's treatment of technical progress as "almost non-existent," recommending that this be given "urgent consideration." He also regarded the model's treatment of uncertainty as "inadequate." In turn, Whyte used Barker's lack of access to the model as a counterargumentative strategy: "This report is interesting because the author broadly endorses what has been done, although on many items he would not have realised this when preparing the Report. . . . By and large it gives strong support to what we have in fact done, albeit in some cases unknown to the author."[60] Whyte effectively suggested that Barker's criticism of the model was an unknowing endorsement of the Post Office. Returning the model to an opaque state allowed Whyte to reconfigure criticism of the model as trust in the Post Office.

The Treasury also reviewed the model and, like Plessey, had to do so by analyzing the Post Office's report. But, unlike Plessey, the Treasury gave its tacit approval. Steve Littlechild, an academic economist who had joined the Treasury as a part-time consultant, reviewed the model.[61] An expert in mathematical programming, Littlechild had previously applied linear programming to analyze US telephone services. He thus chose to analyze the Post Office's decision to choose simulation over linear programming, and advised the Treasury that simulation was "at best a crude technique for finding an optimum" using "hit and miss" repetition to find the best strategy.[62] He concluded, however, "that the Post Office has developed a powerful approach for dealing with their investment analysis" and that he had "no comments."[63] The Treasury passed their report over to the MPT, who interpreted Littlechild's conclusions as an endorsement of the model, writing that "unless we have been seriously misled it seems as if the simulation model has achieved what was required of it and that therefore there would be no substantial advantage in reformulating the model in programming terms."[64] This quote underscores the level of trust that the Treasury placed in the Post Office. Treasury officials did not bemoan their lack of access and implied that, even if the Post Office had misled them a little bit, its economists were confident in the Post Office's approach to simulation. The Post Office's discretion thus received tacit approval from the Treasury, highlighting the trusted opacity of Post Office computing.

The Central Policy Review Staff analyzed the dispute with a broader remit, addressing the modernization model and the Post Office's monopsony over its suppliers. The CPRS, formed in 1970 by Ted Heath, was a cabinet office think tank established to help the cabinet take a long-term policy view.[65] The CPRS thus considered the Post Office's monopsony over its suppliers, which the Industrial Reorganisation Corporation had reorganized in 1968 after the Highgate Wood failure. The IRC recommended that the cooperative arrangements of the Joint Electronic Research Committee, which had developed Highgate Wood, be replaced by an "arm's length" arrangement instead and encouraged greater competition between the manufacturers.[66] The CPRS thus set out to answer three questions: First, would TXE4 be ready by 1975, as the modernization plan had promised? Second, was the Post Office's economic evaluation correct? Third, was the Post Office's strategy best for both the nation and the Post Office?[67]

The CPRS reported that everything appeared on schedule, while somehow also agreeing with GEC and Plessey that the Post Office's plan was "wildly optimistic." On the first question, it thus hedged its bets by endorsing the Post Office's computer simulation but recommending that the Post Office nevertheless prepare contingency orders of Plessey's Crossbar 5005. The CPRS's answer to the second question, about the Post Office's economic evaluation, similarly concluded by endorsing the Post Office's model, saying that it seemed "basically sound," and attributed GEC and Plessey's ire to the Post Office's previous secrecy about the model, followed by its sudden publicity. Finally, on the structure of the telecommunications industry, the CPRS took the view that the "pendulum has swung too far" toward competition, which was undermining collaborative research. The CPRS thus recommended that the Post Office and its manufacturers form a joint development company, with guaranteed funding from all parties and a guaranteed market share for manufacturers at the end of exchange development. The CPRS concluded that the Post Office's modeling approach had been broadly correct and recommended that the Post Office have ultimate design authority over new exchanges. Like the Treasury, the CPRS, in the model's absence, equated the model's authority with the Post Office's authority. By centralizing decision-making with the model and limiting access to it, the Post Office had created a situation wherein the CPRS could not decide whether it approved of the model, but instead must decide whether it approved of the Post Office and its monopsony. In doing so, the Post Office created a situation where the CPRS implicitly supported centralizing all telecom decision-making, including design authority, purchasing, and development, within the Post Office, so reinforcing monopsony.

Finally, as the Post Office's Whitehall counterpart, the Ministry of Posts and Telecommunications also needed convincing. The ministry's reviewers received access to the model, something that no other party had received. The Post Office permitted two ministry economists, H. Christie and D. C. Young, to visit J. S. Whyte.[68] Even then, Whyte carefully managed the meeting. He showed Christie and Young MICES, the Model for Investigating Competing Equipment Strategies, a pared-down version of the original model, ALEM. The meeting highlighted some of the model's limitations and assumptions. Christie and Young criticized the model's conception of supply and demand, as the model did not simulate any interaction between the two and instead assessed the cost of meeting a given demand set by the Post Office. There was also no parameter for capital rationing, so the Post Office assumed it would have all the necessary capital, precluding any future economic difficulties. The absence of capital rationing seems unusual, as the Post Office had suffered investment restrictions after the 1966 July measures, and from 1973 would also suffer pricing restrictions to counter stagflation. Moreover, two other significant British-supported technology projects, Concorde and the advanced gas-cooled nuclear reactor, also experienced difficulties with capital rationing during the 1970s.[69] The Post Office had developed its model, however, in the wake of its corporatization and between the July measures and the Heath government's 1973 Price Commission. The model's optimism thus perhaps reflected the managerialist, technocratic spirit that had informed the Post Office's release from the Civil Service and transformation into a nationalized corporation.

The Post Office followed up on the ministry's review with further defenses. After Christie and Young's visit, Whyte emphasized that "the model does not in any sense make decisions, it merely calculates the consequences of different strategies."[70] By reducing ALEM to a "calculating machine," Whyte emphasized that its quality instead rested on the quality of its inputs and assumptions, which came from the Post Office. In doing so, Whyte again followed the discreet modernist strategy of minimizing computerization and emphasizing executive decision-making. Meanwhile, William Ryland, the Post Office's chairman, wrote to John Eden, invoking further expertise supporting the model. Sir James Lighthill, the renowned physicist, part-time Post Office board member, and author of the infamous "Lighthill report" on artificial intelligence, reviewed the model twice and gave his support.[71] Ryland also referred to similar approaches by Bell of Canada, who had used a similar model, and by AT&T, which, "using a less elaborate model, have acknowledged the greater depth of our approach."[72] Eden was convinced, and his full appraisal for the Cabinet supported the Post Office, citing the numerous other governmental reviews that had also endorsed the Post Office.[73]

Key to the Post Office's success in this controversy was its reliance on discreet modernist strategies, which ensured that the model's authority became conflated with the Post Office's authority, allowing the computer to fade into the background. J. S. Whyte recast Plessey's criticisms as an unknowing endorsement, while the Central Policy Review Staff, in the absence of the Post Office's model, chose to endorse the Post Office's monopsony over its suppliers. The Treasury's response emphasized the trusted opacity of Whyte's model. John Eden's full appraisal of the controversy supported the Post Office's TXE4 plan, citing the various reviews that had endorsed the Post Office and, by extension, Whyte's model.[74] The controversy caused by publicizing the simulation, and its resolution through returning the model to an opaque state, underscores the successful tradition of discreet modernism in British state computing projects. Whyte's career came through this controversy not only unscathed but perhaps even strengthened. After Merriman retired in 1976, Whyte replaced him as the senior director of development, still colloquially known as the engineer-in-chief.[75] But beyond Whyte's career, the ALEM TXE4 controversy shows an early victory for the "self-governing" digital network. Merriman had outlined the need to computerize both the network and its network's management, so developing an evolutionary, resilient infrastructure. This philosophy was inspired by Merriman's time as a Treasury O&M expert mechanizer inside Britain's "government machine," and Whyte, another former Treasury O&M mechanizer, used the discreet modernism of Britain's government machine to defend the Post Office's TXE4 modernization strategy. The computerization of exchange planning thus shows how thoroughly the government machine shaped the Post Office's monopsony over its suppliers.

SYSTEM X: PICKING WINNERS

The last two cases showed how automation and computerization influenced the Post Office's monopsony over labor and supply, respectively. System X, Britain's first computerized digital telephone exchange, launched just as British telecommunications was liberalized, brought change for the monopsonies over both labor and supply. This first section on System X explores how System X development and liberalization influenced BT's monopsony over equipment supply. While the TXE4 controversy had been unfolding, Roy Harris's Advisory Group on Systems Definition had also been proceeding with its goals, set by Merriman, to define and specify a framework with industry for Britain's first computerized digital telephone exchange.[76] Because of difficulties in coordinating the suppliers, it took almost five years for the AGSD to submit its final report. The TXE4 simulation dispute had delayed AGSD negotiations by almost a year, but further delays came from the Post Office's attempts to shake up the equipment supply market. In 1972, as the TXE4 dispute was winding down, the Post Office moved to bring in a new, external supplier, Pye/TMC, owned by the Dutch electronics manufacturer Philips. The Post Office board believed that competitive pressures might bear more on domestic suppliers if Philips established a foothold in Britain.[77] This move exacerbated relations with the Post Office's "court suppliers," GEC, STC, and Plessey, all of whom refused to admit Pye/TMC and undertake AGSD discussions.[78] This resulted in delays for another year, which the Post Office resolved by awarding GEC a contract for the new exchange's central processor. This broke the deadlock, and the Post Office inserted terms and conditions into the contracts for the admission of Pye/TMC, subject to Pye/TMC negotiating its contribution with the other three firms.

The AGSD finally submitted its report in 1972, which laid the framework for a modular family of exchanges collectively named "System X."[79] These exchanges would serve a range of functions within the network, from local to trunk switching. The concept of System X as a modular family of exchanges for local, trunk, and international switching was possibly inspired by IBM. In the early 1960s, IBM had developed System/360, built on a "compatible-family" concept, to maintain software compatibility among different computers. IBM simultaneously developed five computer models for System/360, which was aggressively marketed and announced in 1963 in simultaneous press conferences in sixty-three cities across fourteen countries. System/360 was an enormous success and became IBM's "engine of growth" for the next thirty years.[80] The AGSD's report made no direct mention of IBM or System/360, but given System/360's popularity, the similarity in both systems' names, and the same core principle of software compatibility across multiple modular machines, it seems possible that System/360 inspired System X.

The AGSD's report came around the same time as the Post Office's UK Trunk Task Force report, which used modeling to recommend the digitalization of the entire British telecom infrastructure. The UKTTF's model also informed the plans for System X. Its model recommended organizing System X into "Group Switching Centres," which would centralize traffic across Britain, reducing the number of large regional exchanges from 370 to 200.[81] The combined recommendations of the AGSD and UKTTF established a clear, detailed direction for the development of digital switching. In partnership with industry, the Post Office would develop a modular series of digital exchanges that would serve all different types of switching, but the primary goal was to use these exchanges to build group switching centers that would centralize traffic. In July 1973, the Post Office management board approved these plans for an "all-purpose digital transmission environment."[82] The Post Office created a new department, the Telecommunications System Strategy Department, headed by Harris, to manage System X development. This approach differed from the CPRS's recommendation, discussed above, to create a joint development company with industry to develop System X, which the Post Office had rejected based on the delays that had plagued the AGSD.[83]

Yet this did not help the Post Office escape the delays that emerged from its fraught relationship with its suppliers. Among this new department's main duties was allocating System X development contracts, and in 1975, as it finally prepared to place System X contracts, the Post Office discovered that STC and GEC's cable-making subsidiaries, STC Cables and TCL, had secretly been fixing prices.[84] The Post Office board reacted by ordering an immediate review into the System X contracts and attempted to rewrite them, demanding access to all firms' costs to guard against further price-fixing. Plessey and Pye/TMC, having not engaged in any illicit activity, protested. Moreover, the development contracts were already cost-investigated, so the suppliers would not accept further investigation. Edward Fennessy, the managing director of Telecommunications, persuaded the board to drop this proviso by arguing that, because System X was modular and its components thus interchangeable, the Post Office could also easily interchange contracts. If the Post Office wanted to change supplier for a subsystem, previous subsystems could act as "accurate yardsticks" to measure a contract's legitimacy.[85] System X thus offered a new model for tendering contracts within the monopsony. Just as a new modular subsystem had to be standardized to replace an old subsystem, so the legitimacy of the new contract could be judged by the standards of the old contract.

The Post Office finally placed System X contracts in 1976, but its protracted development had attracted political attention by this point. The

Carter Committee criticized System X's delays, and, inspired by the US liberalization of telecom equipment markets, suggested that the equipment
monopoly in the UK was also suitable for liberalization.[86] This did not yet
happen, but the committee also recommended that the Post Office create a
separate department responsible for liaising with suppliers.[87] The Post Office
thus created a System X Development Department, headed by John Martin,
who had previously supervised systems planning under Harris in the Telecom System Strategy Department.[88] System X development entered the final
stretch, and in 1979 it debuted at the International Switching Symposium in
Geneva. System X debuted domestically in 1980 at the National Exhibition
Centre, and that year the first System X exchange was installed at Baynard
House in the City of London, followed by a local exchange in Woodbridge,
Suffolk. In 1981, two more exchanges were installed, and by 1983 a further
twenty were planned.[89]

The Thatcher government used System X's debut and BT's liberalization to break up the "telecom club" of BT and its "court suppliers." This
happened in two seemingly contradictory episodes: the protectionist reorganization of System X procurement contracts, and the introduction of a
foreign pacesetter, Thorn-Ericsson's System Y, to compete with the domestic
suppliers. The Thatcher government, in its first term, reorganized System
X procurement not to liberalize equipment supply and promote competition, but instead to unify System X manufacturing under a single supplier.
Keith Joseph, Thatcher's secretary of state for industry, argued that this
would allow British electronics manufacturing to prepare for competition
on international markets.[90] These manufacturers were also worried about
their export prospects but, in contrast to Joseph, argued for faster liberalization and privatization to open new markets. D. H. Pitcher, Plessey's managing director, lobbied Number 10 to broaden liberalization to include BT's
monopoly over equipment maintenance, arguing that otherwise, BT could
favor its own equipment over maintaining its competitors' equipment.[91]
Arnold Weinstock, GEC's managing director, pressed Joseph to privatize BT
so that it could escape the public-sector borrowing requirement and access
private finance to fund System X procurement further. Almost a decade
later, Weinstock was still bitter about losing out during the TXE4 modernization process "on the basis of unreliable output from a computer study
of the economic case" and argued that private finance would allow BT to
move past its prior modeling-induced ineptitude.[92]

In the end, the government's desire to rationalize System X manufacturing worked against Weinstock. In May 1982, BT notified GEC, Plessey, and STC that, with the government, it had decided to narrow down System X procurement to just one supplier, which would guarantee that supplier a revenue stream to grow its manufacturing base, resist international competition, and market System X abroad.[93] By September 1982, the government had agreed with BT and the manufacturers that STC would pull out altogether, that Plessey would become the prime System X development contractor, and that Plessey and GEC would split manufacturing until 1985, so that both could prepare for domestic and international competition.[94] System X installations soon picked up, and by 1988 nearly 1,300 new exchanges had been installed.[95] Here, the Thatcher government, along with BT, was effectively "picking winners" rather than promoting competition in the domestic ICT sector.

This arrangement did not last long, however. BT, concerned about GEC and Plessey's ability to manufacture System X in the three-year window, from 1982 to 1985, began to search for an alternative System Y exchange from a foreign manufacturer that could act as a pacesetter for GEC and Plessey.[96] In 1984, BT chose Thorn-Ericsson, the British-based subsidiary of the Swedish telecom manufacturer Ericsson. Despite having its factories based in the UK and having 51 percent British-based ownership, Thorn-Ericsson was still a controversial choice. As part of liberalizing telecommunications, the Thatcher government had created Oftel, the Office for Telecommunications, to regulate and promote competition telecom markets. Oftel did not oppose Thorn-Ericsson's entry into the public switching market, as its goal was to promote competition, but it ruled that BT could use System Y to fill only 20 percent of its public switching needs for three years, between 1987 and 1990. This ruling was intended as a halfway measure to protect GEC and Plessey, while still promoting competition. BT, however, rejected Oftel's ruling, and so undermined Oftel's authority. Not long after, Ericsson divested its Thorn subsidiary entirely, and so the Swedish manufacturer itself became the sole supplier of System Y equipment to BT. Rather than protect national telecom manufacturing, the Thatcher government's apparently contradictory approach to the "telecom club" had opened up Britain's telecom equipment supply market to the world.

This approach, of protecting System X supply at first and then allowing System Y supply later, only seems contradictory, however, if viewed from the domestic suppliers' perspective. The common feature of both cases is that the Thatcher government let BT do what it wanted. In the first case, the government supported and helped redesign System X procurement. In the second case, its pro-competition regulator, Oftel, ruled in favor of opening supply to foreign entrants, and the government itself did not fight when BT opened up those markets harder and faster than Oftel had ruled. This approach fits a broader pattern across the developed world in this period, where governments invested in and directed information technology industries to gain a competitive edge on international markets as they attempted to extract themselves from the financial turbulence of the 1970s and early 1980s.[97] For example, in 1982, the president of France, François Mitterrand, announced an expansion of electronics research and development. The French government announced a nationwide program to connect every household to cable television, which would create jobs and fuel the development of French electronics equipment for export.[98] In Britain, the Hunt Committee in 1982 also recommended a nationwide cable program, and Leon Brittan, the home secretary, called cabling the nation an investment in growth, jobs, and Britain's future.[99] But cabling, which proceeded in a regionalized, liberalized model, was a secondary concern for the Thatcher government compared to its main priority, British Telecom. Rather than focus on telecom manufacturers or cable companies, the Thatcher government helped BT access private finance, by privatizing it, and secure its exchange supply, by both rationalizing the UK-based supply and opening supply to foreign entrants. This strategy was designed to help BT, rather than its suppliers, succeed on international markets and provide stable service to its business customers, especially the financial sector (for further discussion, see chapters 6 and 7). Thatcher's political economy of digitalization was thus not a simple free-market project, but was characterized by a contradictory interventionism that picked one winner: British Telecom.

SYSTEM X: "AUTOMATE OR LIQUIDATE"

Digitalization and automation under Thatcher also meant new attitudes toward labor. In November 1982, Kenneth Baker, the minister for information technology, delivered the Cantor Lectures, titled "The New Information Technology," to the Royal Society for Arts, Manufactures and Commerce as part of the Thatcher government's celebration of 1982 as Information

Technology Year.[100] Sir George Jefferson, BT's chairman, chaired Baker's first lecture, "Information Technology: Industrial and Employment Opportunities." Baker began by outlining the importance of technology to economic growth throughout history and then focused on the importance of information to the modern economy, describing its rise as the "age of *homo bureaucraticus*." Baker advanced the familiar argument that while automation and IT destroyed some jobs, they created others and made the economy more efficient. Baker further emphasized that the faster this happened, the better, concluding his lecture with, "The message is automate or liquidate," which Jefferson described as "both stark and hopeful." Alongside this message, the threat of foreign competition loomed large in Baker's lecture, especially from Japan's microelectronics sector, which inspired "information age" discourses while threatening Western industrial economies' dominance. Perhaps the prime example of this "automate or liquidate" techno-nationalism in Britain came from the automotive industry, with the 1980 launch of the Austin Mini Metro, the successor to the Mini, a classic British icon. British Leyland, the state-owned car manufacturer, launched the Mini Metro with an advert that began with D-Day imagery of Italian, German, Japanese, and French cars rolling off landing craft and onto beaches—but this time, the beaches were British. The advert then cut to British Leyland's automated factory, which unleashed a fleet of Mini Metros that drove through British villages to scare the foreign cars back onto their ships. The Mini Metros then swept past Union flag-waving crowds, villages, and saluting war veterans to perch on the white cliffs of Dover, overseeing the foreigners' retreat, and concluding with the slogan "A British car to beat the world."[101] With both System X and the Mini Metro, the Thatcherist technopolitics of automation drew on a political economy of national champions while also relying on the motif of automating mundane labor.

For System X, this collision of information technology, employment, and Thatcherism meant drastic changes for BT's monopsony over exchange maintenance technicians. Senior management, across various internal communications and information pamphlets, used System X to persuade maintenance staff of the benefits of the deskilling, reskilling, and job losses associated with automation and BT's new competitive environment. In 1957, when rolling out GRACE, the Post Office told its staff that "machines must be servants not masters."[102] With the rollout of System X, BT's management issued a similar statement, telling staff that "computers are still the servants of people, as they have always been."[103] They followed this statement with, "Increasingly, however, machines will take over the more mundane occupations and release the talents of British Telecom staff for more creative and interesting work." In doing so, BT's management implied that jobs were available but only for those talented and creative enough. Staff were encouraged to acclimate themselves to this new environment:

> A modern system will, of course, require a modern attitude from the people who operate it. This will inevitably mean changes in skills and outlook on the part of British Telecom staff, but the system offers fair exchange. The up-to-date equipment is more compact, lighter and healthier to work with and the new "user-friendly" environment it creates will contribute to greater job satisfaction.[104]

Again, the implication was clear. If one could change skills and attitude, then the outlook was good. This was the "fair exchange" for subjecting oneself to the new "modern system," while also reminding staff that they now relied on System X for their job security. System X was even cast as a job creator, as internal BT communications claimed that it was "usually the out-of-date who face unemployment."[105] Management also invoked liberalization, informing staff that their job security would depend on BT's ability to attract and retain new business. System X was the "strongest weapon in our armory as we meet tough commercial competition."[106] Management privately admitted, however, that System X would result in net job losses, with a significant decline in workforce forecast from 1986 onward.[107]

System X meant more than just job losses. It also meant further changes to the nature of exchange maintenance labor. Since the introduction of electronic exchanges with TXE4, exchange maintenance labor had gone through turbulent change.[108] Maintenance of the earlier, electromechanical Strowger exchanges needed manual, mechanical skills, and the strong sonic and visual nature of mechanical switches, which were loud, noisy, and visible, meant that technicians could see and hear faults. Electronic maintenance, on the other hand, was more mental and diagnostic, requiring technicians to recognize and interpret opaque errors and faults indirectly, using diagnostic aids. This led to an increasingly rigid and hierarchical internal labor market. Maintenance supervisors, who had come up through the ranks as Strowger technicians, were not retrained at all for TXE4, but were instead reinforced as "lost managers," working from offices rather than on the exchange floor.[109] Neither were junior Strowger technicians retrained for electronic maintenance. Instead, senior technicians were prioritized for retraining, which segmented the Post Office's internal labor market. Junior technicians became less satisfied with their jobs and their career ambitions, especially as they had no other employment options as maintainers of public switching equipment.

System X, which computerized these electronic exchanges, pushed this segmentation even further, because it took first-line exchange maintenance out of the exchange building altogether.[110] Because computers could be used to remotely diagnose System X faults and execute maintenance programs, System X allowed BT to concentrate both the administration and execution of maintenance into operations and maintenance centers, away from the physical exchange buildings. Maintenance was no longer a specific manual or mental labor skill but was instead integrated into computer-based support systems. For the first time, exchange maintenance was no longer located inside the exchange. Not only did this place greater demands for reskilling on maintenance technicians, but it also drastically reduced the numbers who could reskill in the first place, as BT expected System X to reduce maintenance workforce needs by up to 90 percent. BT made this clear in the literature that it published for System X's debut, where BT explained that System X would free up maintenance workers for "more creative" jobs. This declaration was accompanied by photos of male engineers ready to work, sleeves rolled up, tools at hand, misrepresenting the nature of this new labor altogether.[111]

The changes that System X brought to telephone operator labor, on the other hand, were almost completely invisible and unremarked on. In these same publications about System X, where BT also proclaimed the advantages of digital automated operator services, no mention was made of the mostly female labor force that these services would displace. As with the exchange maintenance technicians, computerization meant changes to both the nature and the numbers of operator labor. In 1983, BT placed a £32 million contract with STC for four thousand computer terminals for directory inquiry operators.[112] The automation of long-distance dialing and computerization of telephone exchanges meant that operators were no longer needed to switch telephone calls, but instead to answer customer queries. As with the maintenance technicians, this ended the colocation of operator labor and exchange buildings, so that operators instead, like the technicians, worked on computer-based support systems. This also meant job cuts, and the computerization of directory inquiries meant that BT expected to reduce the workforce by 25 percent, reducing directory inquiries operators from 10,000 to 7,500.

But, more than this, the computerization of directory inquiries also changed the ways that telecom engineers understood the nature of operator labor. As early as 1972, long-range planners had considered how computerized exchanges could transfer work from operators to customers.[113] This transfer would occur through an interactive process wherein the caller, in response to computerized verbal commands, would undertake a "series of simple acts"—keying or dialing a number, usually a single digit. The sequence of numbers, keyed according to the verbal commands of the exchange, would allow the caller to use services like directory inquiries. In this vision, the computer commanded the caller, as long-range planners described how "the initiative [is] always firmly with the processor." This echoed how Merriman mechanized clerical work while working at the Treasury. Merriman supported computerization by comparing clerical staff's mechanical work "programs" to computer programs, both of which, Merriman argued, worked along discrete, step-by-step lines.[114] This same analogy supported the idea that automated operator services could function by commanding users to follow a series of discrete, simple steps. This debuted with System X as the New Star Services, so called because they used the * button on the keypad. BT explained that customers would use "simple codes" to operate these new services, which included call diversion and three-way calling, and that "an automatic voice guidance system" would provide "step-by-step advice and verification."[115] System X's New Star Services thus realized telecom engineers' understanding of operator labor as a series of simple steps, which a machine could instruct a customer to follow. This stood in stark contrast to the emphasis on the fundamentally human labor of telephone operators in the "friendly telephone" campaign when GRACE debuted. In the 1950s, GRACE had forced feminized adaptations on telephone operators. By the 1980s, System X just forced redundancies.

By the early 1990s, digitalization and privatization had combined to intensify these redundancies among both operators and technicians. In 1990, BT began Project Sovereign, an internal shake-up that restructured the business along business units, rather than geographical regions. Project Sovereign also involved a significant and extended program of redundancies, as BT attempted to become "the world's leanest telecoms company," and so more attractive to international clients and investors.[116] In 1990, BT cut ten thousand jobs, and in 1991, BT announced that it would cut a further 6,500 jobs among telephone operators, explaining that "technology improvements," such as automated directory inquiries, had reduced the demand for telephone operators.[117] Further job losses for operators appeared when BT began to commercialize directory inquiries, introducing charges for customers who used the service. In response, BT operators voted to hold a strike over these job losses, with Alan Tuffin, the general secretary of the Union of Communication Workers, the renamed Union of Postal Workers, arguing, "We cannot allow operators to be treated in this way by a company that is awash with money."[118]

This did nothing to stop Sovereign, which intensified in 1992 as BT planned twenty-four thousand job cuts, offering a redundancy package in April 1992 of three years' salary plus pension for staff that would leave by July 1992. BT also offered three months' work at 25 percent of final salary with job agencies Skillbase and Manpower, and a £1,000 training bond for ex-employees' first year outside BT.[119] By August 1992, twenty thousand staff had taken BT's offer, which the right-wing tabloid *Daily Mail* presented as either "a case of mass optimism or mass foolhardiness."[120] Featuring a photo of the technical and clerical staff who were celebrating redundancy (figure 3.1),one anonymous ex-BT engineer said to the *Daily Mail*, "Something'll turn up. I've a skill. Someone will always want that." This perhaps suggests the difference between technicians' and operators' experiences of how computerization and privatization changed their work. The widespread computerization of technical and engineering jobs meant that technicians now had opportunities elsewhere, whereas for directory inquiries operators, BT was still the only major employer, as its market share vastly outweighed Mercury's. Digitalization had broken BT's monopsony over the maintainers, but not the operators. For BT, which was trying to pivot in a global market from being a national telephone provider, with public service obligations such as free directory inquiries, to an international, commercialized, communications services provider, neither labor pool was as valuable as it had been.

CONCLUSION

Automation and privatization had profound consequences for the subjects of the Post Office and BT's monopsonies. The move to electronic switching from the 1950s through to the 1970s changed the nature of work for both telephone operators and maintenance technicians. GRACE, the robot telephone operator, meant that telephone operators had to redefine their

Daily Mail, Saturday, August 1, 1992

REDUNDANCY OFFER THAT TURNED INTO A MAD SCRAMBLE

Cheque-book optimism: Today it's time to celebrate in the Sir Christopher Wren but joblessness starts tomorrow

FIGURE 3.1

Workers celebrate redundancy at the Sir Christopher Wren pub, near BT's headquarters in St Paul's, London. Source: Levy, Geoffrey. "Why 20,000 People Walked Out of a Job." *Daily Mail*, August 1, 1992. Gale Primary Sources/Associated Newspapers Limited. © Associated Newspapers Limited.

labor in terms of new feminine virtues of grace and amiability, when previously they had been expected to be formal and pious. New forms of electronic maintenance with TXE4 further entrenched the privileged positions of senior maintenance technicians within the Post Office's closed, internal labor market, while junior technicians were denied reskilling opportunities and maintenance supervisors grew more distant from the work they supervised. Computerization initially had similar effects on both labor forces, as their work moved away from the telephone exchanges and toward centralized, computerized support systems, but there were nevertheless clearly different consequences for these gendered labor forces. Women were forced to different defenses of their labor during automation, from articulating new virtues to striking against computerization, while men, who doubtless still suffered during their redundancies, seemed to believe that they would benefit from the transferable skills that computerization had brought to their work. This reinforces a broader narrative from the gender history of computing that, once automation and computing forces women out of work, they face many barriers to reentry, while the "computer boys" remain part of their members-only club.[121] Both groups still had to cope with a drastic reduction in the size of the workforce, which was compounded by Project Sovereign as BT aimed to become leaner for international markets. And so, from 1990 to 1996, BT cut its workforce from 240,000 to 125,000 jobs.[122] The transferability of digital skills and the opening of telecom markets meant that BT eventually lost its monopsony over labor, but this didn't seem to matter to BT's management because, for international markets, exerting monopsonistic power over labor mattered less than cutting the costs of that labor.

Digitalization and liberalization also unsettled the Post Office's and BT's monopsonies over their exchange suppliers. After the Industrial Reorganisation Corporation brought an end to the cartelized bulk supply agreements that the Post Office had with its "court suppliers," telecom management turned to computer simulations to heighten their control over exchange equipment procurement. The ALEM 6 model turned the Post Office's decision to procure TXE4 exchanges from STC into a "matter of fact." Publicizing ALEM's use caused a controversy, but returning to the British bureaucratic tradition of "discreet modernism" assured that various government departments and groups endorsed the Post Office's authority over its suppliers. The decision to develop System X as an interoperable, modular, computerized exchange offered a way to maintain competitive procurement in the wake of the 1975 cable cartel scandal. If a supplier breached a contract with the Post Office, System X's standardized modules meant that the Post Office could more easily turn to a new supplier, although this was easier said than done, as the domestic suppliers' resistance to Pye/TMC's entry showed. This resistance faltered as System X debuted, however, and the Thatcher governments threw their weight behind BT in its disputes with its suppliers. With seemingly contradictory policies of, at first, a protectionist rationalization of System X manufacturing and, then, a liberal opening up of foreign System Y supply, the Thatcher government showed that its main priority was guaranteeing BT's material supply chains so that BT could grow without state investment.

The Post Office and BT's dominance over its suppliers had increased those suppliers' dependence on the national telecom monopoly, all the while undercutting their international prospects. In 1975, the Post Office represented 67 percent of those suppliers' sales, and by 1981 this had increased to 75 percent. Only 5 percent of suppliers' sales went abroad, yet the Post Office, BT, and the government's decisions meant that those manufacturers also lost shares in the domestic market. Once that market had been liberalized, GEC and Plessey also began losing shares of the market in private exchanges, used by business customers for internal communications, to IBM, Ericsson, and ITT.[123] This meant that Britain's domestic telecom manufacturing sector, dominated by GEC and Plessey, faded away. In 1989, GEC bought Plessey in a joint takeover with German electronics manufacturer Siemens and the two companies divided Plessey's assets between them. GEC subsequently divested parts of its business, including its defense arm, Marconi Electronic Systems, through the 1990s, and continued as Marconi Communications until it was bought by Ericsson in 2005.

This chapter shows that the history of public infrastructure technopolitics should not be only about the monopolies that these infrastructures have over the services that they sell, and that it must also be about the monopsonies that they have over the equipment and labor that they buy. Switching is key to understanding this in public telecom infrastructure for two reasons. First, switching occupied a central role in the Post Office's plans for a digital infrastructure. Public telephone exchanges would become the computer control centers that could route all sorts of digital traffic around the UK. Second, telephone exchanges, as the nodes that tie telecom networks together, are the subject of two of the telecom business's most important monopsonies, equipment supply and exchange labor. As this chapter has shown, these roles intersected. It was precisely because exchanges were at the heart of the Post Office's relationship with labor and industry that they took on such a central place in engineers' vision for a digital infrastructure. As Merriman made clear when Harris began drafting plans for Project ADMITS, which would eventually become System X, senior Post Office management wanted a digital exchange architecture that ensured competitive procurement from their suppliers. And, as Merriman himself explained, his definition of communications for this new digital infrastructure was "not only circuitry; it is man management," and that the Post Office's "greatest problem" was the "deployment, use, and, in one sense, control, of manpower."[124] And, as this chapter has shown, womanpower.

What this meant was that privatization and liberalization only reinforced changes in the Post Office and BT's monopsonies that digitalization had already initiated since at least the end of the 1960s. Redundancy and reskilling had been necessary for telephone operators since GRACE and for maintenance technicians since TXE4. System X and Project Sovereign just took this to a new level. Likewise, for the supply monopsony, the Post Office had already used computerization to gain power over its suppliers well before the Thatcher governments supported BT's rationalization and liberalization of the supply industry in the 1980s. But, while public switching was a core part of Britain's new digital infrastructure, and central to understanding the dynamics of domestic telecom labor and manufacturing, it was not the only ingredient for the Post Office's digital vision. The Post Office also needed new, high-bandwidth lines to connect those exchanges to customers and to each other, and new standards that would enable those lines to transmit not just voice traffic, but video and data traffic too. It is to those ingredients that the next chapter turns.

4 WHO KILLED OPTICAL FIBER? DREAMS OF THE INFORMATION HIGHWAY

Every so often, a story does the rounds on British social media of "how Thatcher killed the UK's superfast broadband before it even existed."[1] Originating from an article on *TechRadar* in 2014, the story is told through an interview with Peter Cochrane, BT's chief technology officer from April 1999 to November 2000. The article begins "in the 70s when Dr Cochrane was working as BT's Chief Technology Officer," although Cochrane was actually a research engineer at the Post Office's research center in Suffolk at this time.[2] The story, as *TechRadar* and Cochrane tell it, is that through the 1980s BT began a massive rollout of optical fiber across the country but, in 1990, the Thatcher government decided that this was "anti-competitive" and so BT had to stop. This, Cochrane argues, is why the UK lags behind South Korea, Japan, and much of Europe in rolling out fiber-optic cables to consumers' homes. In reality, this story is more complex than a single decision made by Margaret Thatcher in 1990. The Post Office and BT had been searching for high-bandwidth transmission technologies that would meet the UK's information needs since the late 1960s, when the telecom business first formulated the plan for a nationwide, integrated digital network that could provide voice, data, and video services to customers. It was the long techno-political history of that search that shaped the rollout of optical fiber in the UK, and not a single political decision made in 1990.

This chapter is about that search. It looks at the transmission media—microwaves, coaxial cables, and optical fiber—that telecom engineers thought could provide the UK with the necessary bandwidth for an information society. This is a history of frustration and anticipation. Engineers and managers kept investing in transmission technologies that they thought would meet the demand for an integrated digital network for video, voice, and data, a demand that never materialized and was never seen as politically important. This chapter thus seeks to understand how and why engineers persisted with this anticipatory development, and uses metaphors to help interpret this history.

Metaphors have a long history of structuring thought about technology and society.[3] Machine metaphors of clockwork and computers influenced liberal democratic and clerical bureaucratic modes of government, while network and platform metaphors for political and economic life have become popular more recently.[4] In the nineteenth century, the nervous system and telegraph networks became metaphors for each other, used by scientists and society to better understand electrical transmission as means of bodily and social discipline.[5] Organic metaphors also became popular descriptions of the early AT&T system, described as a "living conscious being" and an "ever-living organism." This presented AT&T's nationwide telecom infrastructure as both holistic and complex, naturalizing a single national system with a single operator, a "careful steward."[6] In the digital era, metaphors have been particularly prevalent in surfacing values and expectations about the internet, whether it has been described as an "information highway," a "digital library," an "electronic marketplace," or a "digital world."[7] These cases all show how metaphors have acted as "midwives," structuring expectations for emerging technologies, yet were also a normative resource deployed by actors to shape the future of these technologies.[8] This chapter thus examines the various metaphors that the Post Office and BT deployed for transmission media, and particularly their use of the "information highway" metaphor, to better interpret how and why the Post Office and BT were developing particular technologies.

But this history is about more than just these transmission media. It is also about the content carried over those transmission media. The Post Office's vision was for an integrated digital network that would carry both data and video, as well as telephony, and engineers believed that this would protect their monopoly from specialist entrants offering only data services. This integration required a standard for digital transmission that would allow a network to transmit all these services simultaneously and interconnect internationally with other digital networks. This chapter thus also explores the development of a new standard, the integrated services digital network. ISDN was favored by the European and Japanese public telecom monopolies and would allow digital networks to simultaneously transmit telephony and data services but, crucially, not television.

Standards—technical rules and conventions that govern how objects and processes are made and used—are vital to technology and infrastructure in modern life.[9] They are important because they embed ideological and political goals in material infrastructure. This means that standards can end up as forms of public policy, and their materiality gives standards an inertia that makes changing them politically and financially costly.[10] Standards have been important for national and international governance since the nineteenth century, helping states manage economies and internationalists create global markets.[11] Since the 1960s and 1970s, a new wave of standardization has emerged in the digital realm. Digital standards have become a contested site for different national and international groups supporting different conventions and politics. During the 1970s and 1980s, a digital standards war emerged between those standards supported by commercial and academic computer scientists and network engineers and those informed by public telecom monopolies.[12] The history of the Post Office and BT's relationship with ISDN standardization thus helps further understand how the vision for an integrated digital network evolved as different technologies emerged and as Britain's telecom infrastructure turned from national monopoly to private operator.

Overall, this chapter thus explores the changing political economy of digital transmission and integration. It investigates the different transmission media and standards that the Post Office and BT hoped would deliver on their vision of a high-bandwidth integrated digital network, and complicates the myth that, if it were not for Thatcher and liberalization, the UK would already have a full fiber-optic network. This history begins in the late 1950s, when the Post Office's focus was on a nationwide network of microwave towers with the Post Office Tower (now BT Tower), London, at its heart. The Post Office Tower served the public not just technologically but also as an instrument of social democracy, and the Post Office's metaphor of the tower as a "lighthouse" combined these functions. The development of this microwave network, which transmitted both telephony and television, reveals the Post Office's disinterest in the early 1960s in integrating these two functions.

This changed with the waveguide, a method for burying extremely high-frequency microwaves underground, which the Post Office pinned its hopes

on during the 1970s as a high-bandwidth "information highway" for integrating voice, television, and data transmission along Britain's long-distance
trunk lines. The chapter then explores how the Post Office's integrated vision
began to falter as the development of the ISDN standard for these integrated
digital networks began. The market for data services had developed faster
than expected, while television remained resistant to integration, and so the
Post Office began to depart from its integrated vision. It remained committed
to entering the television market, however, and so the chapter turns to new
transmission media—coaxial cables and optical fiber—that the Post Office
and BT hoped could still provide "information highways" for the UK. These
plans were frustrated both by government regulation of telecom and media
and the development of the ISDN standard. The chapter thus concludes by
reflecting on how and why digitalization, privatization, and liberalization
meant that, for BT and for the UK, integrating digital services into a high-
bandwidth national network succeeded in some ways and failed in others.

MICROWAVES: LONDON'S LIGHTHOUSE

Microwave transmission started with the exploration of centimeter-
wavelength radio waves, ranging from 300 MHz to 100 GHz frequencies,
as a potential medium in the 1930s. These short waves, first called "micro
waves," were also known as quasi-optical waves because of their similarity to
light. Both traveled in straight lines, lost intensity over distance traveled, and
required direct line of sight. In 1931, ITT's Paris laboratory began experiment-
ing with microwave terminals transmitting telephone and telegraph messages across the English Channel. By 1934, the first Anglo-French microwave
service opened, and in the UK, the Post Office opened a second commercial
link in 1937 between Stranraer, Scotland, and Belfast, Northern Ireland.[13]
World War II, however, interrupted, and not until the 1950s did Britain's
microwave network develop further. The first microwave links to open after
World War II in the UK were for television. In 1949, a London–Birmingham
link opened, followed by a link between Manchester and Kirk O'Shotts,
Scotland, in 1952. In August 1950, British and French engineers conducted
the Calais Experiment, the first live transnational television transmission,
using microwave links to broadcast a variety program, *Calais en Fête*, presented by Richard Dimbleby and filmed in Calais, to BBC viewers.[14] Full-scale
microwave telephony development started only in 1956, as it was harder to

transmit large numbers of telephone calls than a single television signal.[15] By 1964, the Post Office had opened microwave telephone links between Manchester and Newcastle, Elgin and Kirkwall in Scotland, and from Carlisle to Belfast (figure 4.1). Birmingham and London, however, remained isolated because their built environments and natural geography meant that it was challenging to provide line-of-sight links to the city centers.[16]

For this reason, the Post Office had planned two communication towers for London and Birmingham. Work on the Post Office Tower in London began in 1961, but it was not a smooth road from planning to building. Dame Evelyn Sharp, permanent secretary for the Ministry for Housing and Local Government from 1955 to 1966, wrote to Postmaster General Ernest Marples in 1959 to express her and her ministry's "horror" at the plans for such a "particularly conspicuous" tower.[17] Marples defended the tower as a "bold and imaginative solution" to London's line-of-sight problems, endorsed by the Royal Fine Arts Commission, and that it would also offer public amenities in the form of a revolving restaurant and observation gallery.[18] Construction went ahead, finishing in 1964, and the tower opened to the public in 1966.

The Post Office used the tower's amenities and visibility to present itself as a scientific and technological public service. A visitors' information booklet explained how the tower was "a symbol of the modern Post Office, a science based industry using the most refined techniques in the telephone, teleprinter, television and computer communications so necessary for modern society."[19] The presentation of the tower as a high-tech symbol also occurred outside the Post Office. In 1964, the tower appeared in *Eagle and Swift* boys' comic, which played a prominent role in science and technology popularization in 1960s Britain.[20] The Post Office Tower appeared in two recurring features: the adventures of Dan Dare, "Pilot of the Future," and *Eagle and Swift*'s cutaway illustrations of technological wonders, which had previously included examples such as the LNER *Mallard* high-speed steam locomotive and the Avro Vulcan bomber. In *Eagle and Swift*'s May 31, 1964, issue, the front page showed Dan Dare locked in battle with Xel, a dangerous alien, atop the Post Office Tower, while the cutaway illustration inside showed off the tower's interior and exterior workings.[21]

The Post Office Tower also played a key part in the Post Office's role in British social democracy. The Post Office Tower's public revolving restaurant and observation galleries situate the tower alongside other social democratic

Statute miles

25 0 25 50 75 100

Telephony link, already
operational

Telephony link, operational
by 1966

Television link for 405-line
services, already operational

Television link for 625-line
services, operational by 1966

FIGURE 4.1

The British microwave radio-relay network in 1964, showing links already constructed
and those in development. Source: D. G. Jones and P. J. Edwards. "The Post Office Network of Radio-Relay Stations. Part 1—Radio-Relay Links and Network Planning." *Post
Office Electrical Engineers' Journal* 57, no. 3 (1964): 147–155. Reproduced with permission from the Institute of Telecommunications Professionals.

projects such as London's Royal Festival Hall, built in 1951, where publicly accessible spaces could generate new patterns of social relations for a democratic, modern Britain, allowing people from different backgrounds to mingle. In the case of the Post Office Tower, the observation gallery was a place not just to mingle but also to survey the landscape of the nation's capital. In practice, however, the public spaces of the Post Office Tower may have served to reinforce, rather than break down, class differences, as the reservation-only revolving restaurant admitted only citizens wealthy enough to pay its high prices, while the cheaper observation gallery was open to the masses.[22]

The Post Office developed a metaphor of the tower as a "lighthouse" that combined its technological and democratic modernity, interlinking technological progress with public service. The brochure for the tower's opening ceremony described the tower as a "lighthouse-looking structure," while an information booklet for visitors described the tower as a "giant lighthouse."[23] A 1962 press release also described the tower as a "modern, slender lighthouse." In this press release, the metaphor first drew parallels between the revolutions of the restaurant and a lighthouse's signal and, second, linked the tower to an early Victorian telephone exchange, which, built on the roof of a courthouse, was also described as a "lighthouse."[24] So, *pace* Ronald Coase, as lighthouses provide a public service to shipping, the "lighthouse" metaphor here, presenting the tower as both a public amenity and a technological successor to Victorian exchanges, highlighted the Post Office's role in social and technological progress.[25] The "lighthouse" metaphor was also apt because it evoked the quasi-optical nature of microwaves, and, when combined with the connotations of the lighthouse as a navigational instrument of public service progress, suggests the deep entanglement of the tower's threefold function: a node in the microwave relay network, a symbol of technological progress, and a site of democratic modernity.

The Post Office's focus on London's "lighthouse" shows, however, that as much as metaphors emphasize some aspects of technology, they can conceal others. In this case, the lighthouse obscured, rather than illuminated, Britain's nationwide microwave network, as well as the Post Office's other plans for microwaves. The tower was more than just a microwave transmitter and receiver, also housing three telephone exchanges and the London Television Switching Centre.[26] These exchanges, obscured by the tower rising above them, were as crucial to the microwave network as the aerials at the top of the tower yet were much less visible. The tower's television center did not feature heavily, relatively speaking, in publicity. At this point, telephony and television were transmitted via separate, analogue networks, and so the Post Office likely did not yet see television transmission as something that could be integrated into their telecommunications network. This interest would change significantly through the 1970s.

The Birmingham Radio Tower, built at the same time as the Post Office Tower in London, also received less publicity but did further reinforce the "lighthouse" metaphor and the Post Office's commitment to technological modernity. The radio tower featured in the Post Office's "Progress" series of publicity posters (figure 4.2) and, like the Post Office Tower, was compared to "a giant lighthouse but instead of sending out a beam of light it transmits microwave radio signals," further connecting the lighthouse metaphor to microwaves' quasi-optical properties.[27] Even less visible was the nationwide infrastructure of smaller microwave relay towers that linked cities together. By 1966, there were 120 relay towers, situated at intervals of about 30 miles, with an aggregate route length of 2,000 miles.[28] By 1969, the Post Office planned to miniaturize these relay towers even further and introduce them into the local networks. A promotional film, *Telecommunications Services for the 1990s*, made by staff at Dollis Hill, the Post Office research station, described how microwave poles on every street corner would beam voice, data, and video signals into residential homes.[29] By the end of the 1960s, microwaves were thus beginning to feature in the Post Office engineers' visions for an integrated digital network that would combine telecommunications and television.

While the Post Office's plans for microwaves would continue, the Post Office Tower's fate was less hopeful, instead illustrating the passing of the Post Office's moment of democratic modernity. On October 31, 1971, a bomb exploded in the tower's restaurant toilets. At first attributed to the Provisional IRA, the bomb has since been linked to the Angry Brigade, a British anarchist collective active from 1970 to 1972. The tower subsequently closed to the public and, after the restaurant's lease expired in 1980, the restaurant remained closed.[30] With privatization, the tower became BT Tower, and in the 1990s BT turned it into a space for corporate functions. This turn from public to private did not go unnoticed. In 1994, the *Independent* newspaper criticized "the ephemeral fizz of public relations receptions" that excluded the public, and a 1995 article in the *Scotsman*, "Tower to the People," also laid in, asking, "How come, 30 years on, nobody except corporate fat-cats can

The 500 ft. high Post Office Tower in Birmingham will be a key radio station in the national communication network. It will carry telephone and television systems to and from most parts of the country.

GPO

FIGURE 4.2
Birmingham Radio Tower also featured in the "Progress" poster series (as in chapter 2, for the Post Office's Computer Centre). Source: TCB 420/IRP (PR) 6, BT Archives. Courtesy of BT Group Archives.

get inside it?"[31] The tower continued to transmit into the 2000s, giving it a healthy forty-year lifespan, but by the 1990s, it was clear that the Post Office Tower's era of democratic modernity was over.

WAVEGUIDES: SUPER COMMUNICATIONS HIGHWAY

Even as microwaves crossed Britain's skies in the 1960s, the Post Office's attention had already turned to a new earthly transmission medium, milli-metric waveguides, which are metal tubes that direct millimetric microwaves. Millimetric microwaves have shorter wavelengths than the centimetric microwaves used for aerial microwave transmission. The shorter wavelength accentuated microwaves' quasi-optical nature so that, while centimetric microwaves could propagate through air, millimetric microwaves degraded too much during rainstorms and other adverse conditions. The trade-off, however, was that as wavelength decreased, frequency increased, which meant higher bandwidth and greater transmission capacity. The Post Office thus turned to waveguides as a housing and guidance system for millimetric microwaves. In the UK, the Institute of Electrical Engineers held the first conference on millimetric waveguides in 1959, and a conference report in the *Post Office Electrical Engineers' Journal* suggested that further research and development work was needed.[32]

The Post Office was not the only organization to explore waveguide development. Bell Labs also undertook a lengthy research and development program on waveguides, which was abandoned by 1973, and Standard Tele-communications Laboratories also conducted some early research into wave-guides.[33] Alec Reeves, the creator of digital pulse-code modulation, was the chief figure at STL assessing waveguides and was unconvinced, particularly regarding limitations in waveguide geometry. To avoid signal degradation, waveguides needed to avoid kinks and tight turns. The significant distances between urban centers in the US meant that waveguides, requiring gentle, sweeping curves, potentially suited AT&T's needs. Reeves, however, was skep-tical that waveguides suited Britain's densely built-up landscape, so, in 1963, STL abandoned their waveguide project to pursue optical fiber transmission instead.[34] The Post Office continued with waveguide research, expecting that its integrated digital network would come to fruition soon enough that it needed to find a high-bandwidth backbone for Britain's digital infrastructure straightaway.

By 1967, engineers began field trials at Martlesham Heath, the location for the Post Office's new research center. Engineers began with a one-mile circular waveguide test, working with Professor Harold Barlow, professor of electrical engineering at University College London.[35] By 1975, the Post Office had begun a 14 kilometer trial between Martlesham Heath and Wickham Market, a nearby town. The trial revealed the need for significant auxiliary infrastructure, including a special bridge to navigate a river, a purpose-made tunnel for a stream and marsh, and "mirror corners," using reflective surfaces to bounce the beams at tight angles, in segments with unavoidable sharp bends. Engineers also had to install mechanical tensioning equipment at both ends to hold the waveguide taut, mitigating soil subsidence and overcoming expansion and contraction during temperature changes.[36] Combined with STL and AT&T's abandonment of waveguide research, these issues invite the question of why the Post Office continued with the waveguide. There were several reasons. Both the Post Office's vision for an integrated high-speed digital network and its expectations of an impending information revolution meant it predicted great demand for high bandwidth in the near future. Furthermore, the Post Office also expected that optical fiber, a high-bandwidth alternative, would take much longer to develop than waveguides.

Showing again the power of long-range planning in the Post Office, the UK Trunk Task Force's study on the integrated digital network, run by the Long Range Studies Division, strongly influenced the Post Office's commitment to the waveguide. The UKTTF's modeling studies, which affirmed the digitalization of the entire network, also affirmed the waveguide as the natural next step for transmission. The UKTTF highlighted the waveguide as a technology suited to both the increasing traffic that the Post Office expected and the multimedia nature of that traffic, including voice, data, and video. These expectations and forecasts may have been reinforced by the fact that, when Harris first started establishing the requirements for the general-purpose digital network in "Telecommunications Systems of the 1980s" (discussed in chapter 2), the press frequently criticized the Post Office for not keeping up with public demand for telephones, and in the early 1970s, when the UKTTF made these forecasts, this demand sharply increased (figure 4.3).[37] The UKTTF report showed that digitalization was economical regardless of transmission medium, but the waveguide was the most cost-effective and high-bandwidth transmission medium available. The task force's final report included a recommended waveguide layout for 1986, stretching from

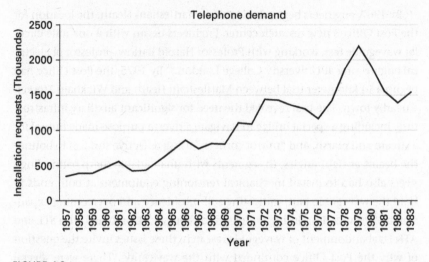

FIGURE 4.3

Public demand for telephone installations from 1957 to 1983. As well as increasing from 1967 to 1972 and from 1976 to 1979, demand also slumped following the 1962–1963 capital restrictions, Harold Wilson's 1966 July measures, and during the 1973–1975 and 1980–1981 recessions. Sources: Post Office and BT Annual Reports and Accounts, 1959–1984, TCB 10, TCC 11, TCD 12, TCE 13, BT Archives.

London to Bristol, and Carlisle via Manchester and Leeds, with onward connections to Cardiff and Glasgow if there was enough demand for Viewphone, which again appeared as an avatar of Britain's integrated digital network.[38] The UKTTF report also, however, marginalized optical fiber as a potential high-bandwidth digital transmission medium. The task force viewed optical fiber as an experimental technology and did not include it in any forecasts.[39]

The relationship between optical fiber and waveguides continued to influence appraisals of both media through the 1970s. In 1972, a report of the meeting of the Managing Director's Committee for Telecommunications, including James Merriman and the managing director Edward Fennessy, noted that optical fiber research showed greater potential than expected but predicted that the trunk network would still use waveguides, while optical fiber was better suited to the local network.[40] In the mid-1970s, Post Office publicity and information brochures described waveguides as the "big brother" of optical fiber, carrying heavy communications traffic, whereas optical fiber would serve local networks.[41] This complementary attitude was prominent at a 1976 IEE conference on the millimetric waveguide. In his opening address, Merriman argued that waveguides "will have to be judged—against their timeliness and relevance or irrelevance—to the spectrum of competing or complementary technologies."[42] Merriman reviewed potential alternatives to the waveguide: coaxial cable, low-bandwidth but already in extensive use; microwaves, already exhausting the available frequency spectrum and vulnerable to degradation during heavy weather; satellites, economically viable only for international transmission; and optical fiber, "not yet in a position to compete" in the trunk network, but a promising option for local networks. Merriman concluded that waveguides were the clear victor, referencing "their extraordinary capability for information/ bandwidth."[43]

The waveguide's primacy peaked in 1977 when the Post Office board approved Britain's first trunk transmission system designed specifically for digital use, a link between Reading and Bristol, connecting South Wales and the West with London. Bristol was also an important node for routing international traffic to London from submarine cable stations in Cornwall and the Post Office's two satellite earth stations, Goonhilly in Cornwall and Madley in Herefordshire. The telecom business proposed two versions of the link to the Post Office board. The first version, a waveguide system, would be ready by the end of the decade, while the second, an optical fiber link, would not be available until 1983. The Managing Director's Committee for Telecommunications recommended the waveguide system because optical fiber had not yet entered development, and also because the committee believed that the Post Office, as one of the few remaining waveguide pioneers, had the opportunity to showcase its world-leading position and create an export market for the UK. The telecom business's proposal also further showed its high expectations for a future integrated digital network. The proposal explained that while the Bristol–Reading link would use only 20 percent of the waveguide's capacity over twenty years, this meant that there was excess capacity for other content, including videoconferencing.[44] The Post Office's management board approved the waveguide in February 1977, not least because its members also saw the Bristol–Reading waveguide as "insurance" against delays in optical fiber development.[45]

The Bristol–Reading waveguide project failed, which signaled the end for the waveguide. The Post Office board, as part of the project's "insurance policy" status, had decided in advance that Bristol–Reading would be the only waveguide link, expecting that optical fiber would be developed before other trunk routes needed waveguides. When Marconi, the Post Office's manufacturing partner on the project, heard the news, it decided to recoup all its waveguide R&D expenditure from the Bristol–Reading link by raising costs. Britain's economic downturn in the late 1970s also affected the project, as telephone growth slowed relative to the UKTTF's earlier forecasts. Long-range planners' optimistic projections of a high-capacity integrated digital network serving telephony, data, and video over waveguides had become excessive after economic slumps slowed telephone use.[46] The combination of slow telephone growth, faster-than-anticipated optical fiber development, and higher manufacturing costs meant that the Post Office canceled the Bristol–Reading waveguide project in 1978, only a year after its approval.

The waveguide was, however, a metaphorical success as the subject for the Post Office's first uses of the "highway" metaphor. According to *Wired Style: Principles of English Usage in the Digital Age*, a 1997 dictionary published by the digital utopian magazine *Wired*, Al Gore Jr., the US senator and later vice president, coined the metaphor "information superhighway" in 1978 to refer to a range of high-bandwidth digital communication technologies. Gore later popularized the term during the 1990s to promote a national information infrastructure when he was vice president to President Bill Clinton.[47] The "highway" metaphor, however, has a much longer history with communication networks. In the 1920s, AT&T used the metaphor "a highway of communication" to compare its telephone network, a government-sanctioned private monopoly, to the US public highway system, attempting to persuade its customers of "the logic and beneficence of a unified system."[48] In the 1970s, the Post Office used the "highway" metaphor to communicate the waveguide's high bandwidth and the telecom business's plans for a general-purpose digital information network. The highway metaphor first appeared in the *Post Office Telecommunications Journal* in 1970, which described waveguides as "highways of communication" and "super-highways for telecommunications traffic."[49] This usage continued throughout the 1970s in publicity about the Post Office's various waveguide projects. A press release for the 1975 Martlesham–Wickham Market trial described waveguides as a "super-highway" solution that the Post Office had prepared for "major telecommunication highways," while press notices for the Bristol–Reading link called the waveguide a "super communications highway" that would carry data, television, videophone, and videoconferencing services as part of an integrated digital network.[50]

The interesting feature of the Post Office's "highway" metaphor is how it employs one infrastructure as a metaphor for another. There is a long history of using metaphors for infrastructure, such as the nervous system for the telegraph, but the highway metaphor is different because it used one national infrastructure as a metaphor for another, rather than applying social or biological metaphors to technology or vice versa. There is a longer history here, too. Telegraphy was used as a metaphor to understand the emergence of radio, which became "wireless telegraphy," focusing early attention on radio's point-to-point communications applications rather than its potential for broadcasting.[51] This suggests that, for engineers, some infrastructures become "paradigmatic" in terms of setting their expectations and defining their assumptions how about their own infrastructures will develop.[52] For the Post Office, the highway metaphor in the 1970s served different purposes from AT&T's use in the 1920s. Where AT&T had used it to defend its private monopoly, the Post Office used it to conjure an image of the future, one in which communications routes would have the high capacity of highways, distinct from local roads, and would carry the mixed "traffic" of data, telephony, and television. While the waveguide failed as a technology, like Viewphone, it successfully reinforced the "highway" metaphor's implication that Britain's future digital network would integrate telecommunications, both voice and data, with television and video services. But Post Office engineers and managers soon began to develop new digital ambitions that pushed the integrated vision beyond its breaking point. They simultaneously tried to expand into television transmission, contributed to integrated digital standards development, and started building their own specialized data networks, previously anathema to their vision of integration.

STANDARDS AND SPECIALIZATION IN DATA NETWORKS

Chapter 1 explored how one of the key parts of Post Office engineers' vision for Britain's digital future was an integrated digital network that combined voice, data, and video services, and that one of the motivations behind that vision was to defend the Post Office's monopoly from specialized, commercial data networks. This vision developed from national experiments with integrated digital networks during the 1960s, such as the Post Office's 1967 trial integrated digital network in Washington New Town, then expanded to further experiments in 1968, and also drew on the successful digitalization of switching networks in Japan, and France.[53] By the start of the 1970s, it was clear not just in the UK but around the world that integrated digital networks were one direction that digitalization could take. The other direction was building separate digital networks for telephony, data, and perhaps even television and other video services such as videoconferencing. But which direction was preferable, and how that vision might take shape, was a question for more than just Post Office engineers. It was a question that, from the start, was transnational, framed and answered by the International Telecommunication Union's standard-setting organization, the Consultative Committee for International Telegraphy and Telephony (CCITT). In June 1971, the CCITT established a sub-working party to discuss the terms "integrated digital network" and "integrated services network," which led to the term "integrated services digital network." The following year, the CCITT distributed a circular letter to national representatives, including the Post Office and other public telecom monopolies in Europe, asking for their opinions on how integrated digital networks should be studied during CCITT's 1973–1976 study period.[54]

The CCITT circular letter offered three options: a "service integrated digital network," as the Post Office had envisioned; "specialised digital networks"; and two specialized networks, one for data and one for voice, with "limited interconnection." The different responses to this question reveal the different ambitions and pressures shaping national telecom operators at the time. Japan lamented that the CCITT question already excluded video services from the ISDN and argued that video would be "indispensable" to the "ideal communication network for the future." Many European PTTs, especially in the Netherlands, Spain, Italy, and France, argued that maximum integration was the best solution, as the most economic and flexible option, and that the national PTT administrations were best-placed to coordinate these integrated digital networks. The US, on the other hand, argued against integration and in favor of separate digital networks for telephony and data, asserting that they were sufficiently different services that they required separate networks, and that an ISDN could delay establishing specialized data networks in countries around the world.

US opposition to ISDN was a familiar sight from digital standards conflicts that took place during the 1970s, which exposed the vested interests of European PTTs, US computer and data services corporations, and academic computer scientists and network engineers. During the 1970s, the CCITT and

ISO, the International Organization for Standardization, became involved in two particularly fraught battles over the standards for packet switching and the architecture of these data networks. For packet switching, the choice was between datagrams and "virtual circuits." Datagrams built on packet switching's potential to have data messages disassembled by the sending computer, sent by different routes over the network, and then reassembled by the receiving computer. Virtual circuits, which took shape in the X.25 standard, meant that the network would open a virtual circuit between sending and receiving computers, akin to the physical circuit opened between telephone callers, allowing a continuous stream of data packets via a single route. Advocates of liberalized, open, decentralized networks supported datagrams, which shifted control to users' computers, while the European telecom monopolies, including the Post Office, supported X.25, which was modeled on their circuit-switched telephone networks and preserved control within the network.[55] By 1976, the CCITT, dominated by European PTTs, approved the X.25 standard.

For network architecture, the conflict was between IBM's proprietary standard, the Systems Network Architecture, the ISO's Open Systems Interconnection (OSI) model, and the internet's framework, TCP/IP.[56] Standardizers at the ISO conceived of OSI as a critique of both IBM's dominance over the computer industry and of the European PTTs' monopolies over their networks, and therefore called for open standards that would promote opportunities for competitors. ISO's openness meant, however, that IBM and the European PTTs joined the negotiation table, slowing down the standardization process to the point that OSI failed and the internet's TCP/IP framework became the dominant framework for connecting digital data networks around the world. This history shows that US interests were often opposed to European interests in setting standards for digital networks. Standards that preserved European PTTs' control over communications networks reduced opportunities for American computer services and data networking firms to set up specialized data networks in Europe. For the European PTTs, integration and ISDN offered a way to protect their monopolies against competition from these data networking companies.

Most European PTTs thus answered the CCITT in favor of the ISDN, but the Post Office, surprisingly, reversed its position and argued against the integrated digital network. The Post Office instead argued that there was an "immediate and pressing need" for new data communications and telex services, while there was less certainty that digital telephone networks would be rolled out within the decade.[57] The Post Office also believed that an integrated digital services network would be optimized for digital telephony in ways that were not optimal for data services. Submitting a flow chart explaining its argument (figure 4.4), the Post Office concluded that the most logical choice was the second option, developing separate, but interconnectable, digital networks for data and telephony. This seems completely at odds with the Post Office's goal, first outlined in 1967 by Merriman and Harris, of an integrated digital network. These goals had since been repeated and reinforced. In 1971, Merriman had warned the Post Office board that closed packet-switched data networks threatened the Post Office's monopolistic goal of "universal integrated networks," and in July 1973, the Post Office management board, following the recommendations of the long-range planners' UK Trunk Task Force and Roy Harris's Advisory Group on Systems Definition, approved the telecom business's plans for an "all-purpose digital transmission environment."[58]

Two factors might have caused the Post Office's change of heart and its opposition to transnational ISDN standardization. The first, more speculative, reason is that ISDN standardization, as Japan had lamented, omitted video services. From the start, both Merriman and Harris had emphasized that the future digital network would carry not just voice and data, but also video. This view is underscored by the lengths the Post Office went to keep developing the Viewphone and by the Post Office's commitment to the waveguide as a microwave backbone that could carry, like the airborne microwaves that preceded it, both telecommunications and television signals. The Post Office was thus perhaps less interested in a standardization effort that, from the start, excluded video services, and this is reinforced by the fact that, as the next sections show, the Post Office kept investing in technologies and lobbying the government in order to expand into television.

The second, firmer, reason was that the growing domestic pressure for data networks undermined integration. By the early 1970s, it was increasingly clear that the growing demand for data services from British business users meant that the Post Office would have to develop a specialized packet-switched data network. The Post Office thus began developing, in partnership with Ferranti and the British academic community, a domestic data network, which debuted in London and Manchester in 1975.[59] This network, the Experimental Packet Switched Service, aka EPSS, was hugely oversubscribed, and the Managing Director's Committee for Telecommunications

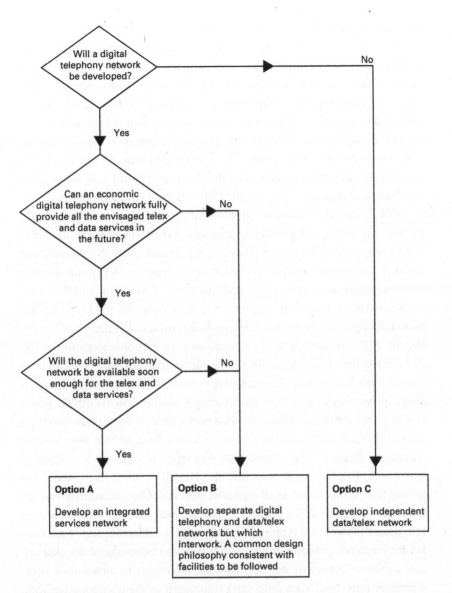

FIGURE 4.4

The Post Office's reasoning against integration for CCITT. Source: The International Telegraph and Telephone Consultative Committee: Fifth Plenary Assembly: Green Book, Volume III-3: Line Transmission. Geneva: International Telecommunication Union, 1973. Reproduced with permission from the ITU.

was particularly concerned that the business sector might lobby the government to liberalize the Post Office's telecom monopoly if it did not roll out
further packet-switched data services.[60] As it happened, the financial sector
lobbied along exactly these lines, so this concern was not baseless (see chapter 7). The Post Office thus continued to develop a packet-switched data network and so, as senior management began to worry that liberalization would
happen regardless, the monopolistic promise of integrated digital services
networks began to lose its allure. The Post Office's retreat from ISDN standardization can thus be explained by the growing domestic business pressure
for specialized data networks and the ISDN's neglect of video services.

ISDN studies nevertheless continued under the auspices of the CCITT
through the 1970s, and formal development of the ISDN began in the 1981–
1984 study period, which set the basic framework of ISDN concepts and
network architecture. Nippon Telegraph and Telephone (NTT), the Japanese
public telecom monopoly, pioneered this effort, developing an ISDN technical concept in 1982 and implementing it in Tokyo in 1984. During this
period, European PTTs also focused more effort on standardizing and launching the ISDN, influenced by the liberalization of BT's monopoly in the UK.
PTTs on mainland Europe saw ISDN standardization as a method to preserve
control over their monopolies against the threat of liberalization.[61] This took
shape in two ways: First, PTTs' multinational business clients usually leased
telecom lines for their communications needs. One of the earliest targets for
telecom liberalization was the market in leased lines, so that new telecom
companies could provide leased lines to corporate customers, competing
with the PTT's own leased lines services. The ISDN, however, would make
leasing lines unnecessary, as all digital services could be transmitted over the
same network—the PTT's network.[62] The second way that ISDN benefited
European PTTs was that it protected them from the liberalization of the market for customer premises equipment.[63] This had been one of the first targets for liberalization in the UK, allowing customers to attach their own,
separately purchased, data processing equipment to the telephone network.
ISDN, however, meant that PTTs could offer their own remote data processing
services over the network, meaning that customers had less incentive to buy
their own data processing equipment. This in turn explains why US information services companies and computer companies, such as IBM, lobbied
in Europe for telecom deregulation. Liberalization meant that public ISDNs
would be more difficult to roll out and would provide market opportunities

for these US firms to offer services and hardware for data transmission and processing.[64]

In the UK, where telecommunications had been liberalized since 1981, and where newly created BT had already been diversifying into specialized data services, the ISDN thus became less important. By 1981, after significant lobbying from business users in the City of London, BT had finally launched its fully-fledged packet-switched data network, PSS.[65] Alongside PSS, BT launched a range of data services called X-Stream Services. BT explained that the "X," connoting its System X digital exchange, signified "*digital* services," while "stream" indicated "the flow of digital information through the network."[66] X-Stream comprised KiloStream and MegaStream, low-capacity and high-capacity data services provided using rented private networks. Another service, SatStream (explored in more detail in chapter 7), opened in 1984 using satellites.[67] BT also rebranded PSS as SwitchStream One to align with X-Stream marketing. In 1984, BT finally launched an ISDN trial, marketed both as Integrated Digital Access (IDA) and SwitchStream Two, which at long last achieved the business's goal of providing telephony and data over the circuit-switched telephone network using System X.[68] But, as liberalization had already happened, ISDN no longer occupied pride of place for BT. Having read the cards ten years earlier, telecom management had long since turned to prioritizing its business customers with specialized services. This turn is explored in more detail in chapter 7, but for now this chapter stays with Britain's transmission infrastructure to explore the Post Office and BT's last remaining hope for a general-purpose transmission infrastructure: building a network that could expand into television transmission.

COAXIAL: CABLE TV

The Post Office's interest in providing television, along with data and telephony, over cable networks stretched back to its early integrated digital network trials in the 1960s. Cable television first appeared in the UK in 1951 as a method to extend television coverage to areas with poor reception.[69] In the 1960s, telecom engineers found that advanced cables could provide up to nine television channels, along with voice and data, over their networks. At that time, however, unfavorable economics meant that the Post Office used a lower bandwidth cable for its integrated digital trials in Washington, Irvine, Craigavon, and Milton Keynes.[70] In the mid-1970s, the Annan Committee, a review of broadcasting in the UK, reignited the Post Office's interest in cable television. Harold Wilson's Labour government convened the Annan Committee in 1974 to review the creation of a fourth and final terrestrial TV channel. By this point, the BBC and Independent Television Authority were using the remaining available TV broadcasting bandwidth. The Annan Committee had only one technical expert, Professor Geoffrey Sims, head of the Electronics Department at Southampton University, and so the Managing Director's Committee for Telecommunications saw an opportunity, under the guise of technical advice, to lobby for a Post Office cable television network.

Post Office management interpreted the Annan Committee's investigation into broadcasting as an opportunity to expand cable television, believing that, technically and legally, the Post Office had the right to develop and administer a national cable network for television, telephony, and data.[71] The Post Office viewed itself as the network provider, leaving "operation," meaning broadcasting and programming, to broadcasting corporations like the BBC and ITA. Merriman's evidence to the Annan Committee, submitted on behalf of the Post Office, outlined its previous experience with cable provision in its early integrated digital network trials. Merriman argued that "the transmission of information was PO business," viewing a national cable television network as a means for the "integration of a wide range of service options in the PO telecomms network."[72] Merriman summarized the Post Office's evidence with the argument that "present and future telecoms services, together with TV broadcasts, could most economically and conveniently be carried on a single wideband network provided by a single administration," which aligned with the Post Office's existing vision of an integrated digital network. The Annan Committee elected not to review cable television, but, apart from its main recommendation to set up an independent fourth broadcast channel, which became Channel 4 in 1980, also recommended that the Post Office provide any potential future national cable television network.[73]

This recommendation remained the case until a flurry of cable television reports under the Thatcher government in 1982 and 1983. The first report was undertaken by the Cabinet Office's Information Technology Advisory Panel (ITAP). The BT board again lobbied the panel, keen to establish a broadband cable network for television, telephony, and interactive information services for banking, advertising, and entertainment.[74] In the short term, BT envisioned using coaxial cable, but in the medium term planned to use optical fiber. The BT board was aware that the Thatcher government's liberalizing stance meant it was doubtful that the ITAP report would recommend creating a public national cable infrastructure, and so the board also accepted that, at the very least, a private venture partnership would be necessary. The board also tried a tit-for-tat approach, suggesting that it would be much more amenable to privatization, which was still not set in stone, if it meant in return that BT could develop a national cable network.

The ITAP report, released in 1982, did not favor BT, instead recommending the expansion and liberalization of cable TV so that customers could receive multiple channels and interactive services from private regional cable TV operators. The report recommended that BT have only a limited role in setting technical standards, explicitly rejecting BT's proposal for a publicly funded national network.[75] The ITAP report was, however, only advisory, and so BT lobbied the government again as a second official review into cable television, the Hunt Committee, began. BT board members particularly criticized that the ITAP report collapsed cable providers and cable operators into one group, arguing that the government needed to consider cable TV in the broader context of a national information infrastructure:

> Broadband links provided initially for entertainment TV should be seen in the context of a national strategy for developing information technology infrastructure; network configuration and technologies should be adopted which were capable of carrying broadcast TV distribution, TV narrow casting, a wide range of interactive services, two-way video switched services and interconnection with national and international telecommunication networks.[76]

Again, however, the report ignored BT. The Hunt Committee's report also did not distinguish between cable provision and operation, meaning that, to the BT board's lament, "the concept of a totally integrated system would not be achievable."[77] The review also ignored BT's argument that a national cable network could provide telecommunications as well as television, instead recommending that the government franchise out regional cable TV systems. As compensation, the report recommended that cable TV franchises not provide telephony, but it left the door open for data provision. BT tried to point out to the government, however, that this latter recommendation was nonsensical, as digital encoding meant that it was impossible to differentiate between voice and data.[78] The government formalized the Hunt Committee's recommendations, which effectively repeated the ITAP report, in a 1983 government white paper, *The Development of Cable Systems and Services*.[79] Cable networks rolled out as regional franchises, but BT and Mercury's duopoly over voice and data was preserved, restricting the cable networks to television. The government permitted BT to participate in consortia bids for cable franchises, and BT consortia successfully secured five: Aberdeen, Coventry, Ulster, Merseyside, and Westminster.[80]

HIGHWAYS: OPTICAL FIBER

In the background, while the Post Office explored waveguides and coaxial cable networks as solutions for its integrated digital vision, researchers at the Post Office and around the world were hard at work developing the next high-bandwidth transmission medium. Optical fiber transmission uses laser-generated light waves to transmit communication signals down ultratransparent glass fibers, which guide light waves through internal reflections from their interior surface. Optical fiber was the next step for high-bandwidth transmission because it uses near-visible light, which, because it occupies a higher frequency band in the electromagnetic spectrum, could carry even more information than millimetric microwaves. While optical fibers look quite different to millimetric waveguides, they thus work on the same principle of guiding electromagnetic waves with tubes.

Standard Telecommunications Laboratories, in Harlow, Essex, pioneered optical fiber in the 1960s. The key figure at STL was Charles Kao, a Chinese-born engineer who, in 1969, created ultra-transparent glass and proved its viability for low-loss laser communications, and later won a Nobel Prize in Physics in 2009 for this breakthrough. The Post Office had a significant influence on STL's research, as STL's parent company, STC, was a manufacturer and not a network provider, and so needed partners and customers. The Post Office, as both a domestic telecom operator and research organization, was an ideal partner. To complement the envisioned trunk waveguide network, the Post Office wanted a local transmission system with a maximum signal degradation of twenty decibels per kilometer, which set Kao's targets for his research.[81] Post Office research staff collaborated with STL, working on new ways to synthesize glass fibers and setting benchmarks for optical fiber's credibility. For example, Post Office researchers verified the purity of glass fibers synthesized by Corning Glass Works in 1970, another breakthrough following Kao's proof of ultra-transparent glass.[82]

These breakthroughs triggered a race through the 1970s and 1980s to create practical optical fiber. Laboratories worldwide created new fabrication techniques for optical fiber and developed small, reliable lasers so that, by 1977, the Post Office had two working trials, a 13km low-bandwidth 8.448 Mbit/s link and a high-bandwidth 7.25km 140 Mbit/s link.[83] In 1979, the Post Office placed orders with STC, GEC, and Plessey to begin fiber-optic installation in fifteen routes across Britain.[84] Several routes tested experimental low-bandwidth links in different environments, such as Welsh lakebeds, but three links were high-bandwidth 140 Mbit/s cables, including a Reading–London link. Only a year after the Post Office canceled the Bristol–Reading waveguide, preliminary tests took place to standardize optical fiber systems. By the 1980s, BT was breaking fiber-optic world records. In 1982, research staff tested a 140 Mbit/s link over 102km, the longest in the world, and BT completed a London–Birmingham fiber-optic link.[85] The following year, BT placed the first order in the world for a commercial fiber-optic link, running between Milton Keynes and Luton.[86] In 1985, another record followed, the fastest transmission rate ever achieved over optical fiber, 2.4 Gbit/s over 32km of cable, and BT research staff received a Queen's Award for Technological Achievement.[87] The late 1970s and early 1980s have been called the "Fiber-Optic Performance Olympics," with records frequently set and broken worldwide.[88] In the UK, these breakthroughs lent additional weight to the Post Office board's decision to cancel the Bristol–Reading waveguide, effectively canceling the waveguide altogether. Through most of the 1970s, the waveguide had been the "information highway," the "big brother" to optical fiber. But, just a decade after Charles Kao's 1969 breakthrough, the Post Office installed its first trunk network fiber-optic trials, and only four years later, BT was ordering and installing commercial systems.

As BT rolled out optical fiber throughout the 1980s, hope remained for a nationwide integrated digital network, coupled to a new metaphor of an optical fiber "national grid." The national grid was an evocative analogy with the United Kingdom's power network, known as the national grid for much of the twentieth century, and attempted to draw parallels with the near-universal connectivity of the power network, as well as its public ownership.[89] Where highways have often stood as symbols of individual freedom and mobility, grids are a communal infrastructure that bind users together.[90] Several government reports in the mid- and late 1980s addressed the idea of a national fiber-optic "grid." In 1986, the Peacock Committee, ostensibly reviewing the BBC's financing, recommended that the government permit BT to construct a fiber-optic national grid for TV and telecommunications.

In 1988, two further reports addressed this recommendation. The first, *Opto-electronics: Building on our Investment*, a report by the Advisory Committee on Science and Technology (ACOST), argued in favor of a state-funded national grid, while the second, *The Infrastructure for Tomorrow*, an official policy report from the Department of Trade and Industry (DTI), concluded that the government should not fund a national grid.[91]

The DTI's main reasoning was that funding BT's national grid would harm competition, but the report developed an interesting secondary argument about the relationship between the fiber-optic national grid and the ISDN digital standard. The report argued that infrastructure policy should not only take into account competition between firms, such as BT and Mercury, but that it should also consider competition between technologies. The government should thus not "pin its colours" on a particular technology.[92] The report, turning BT's ambition for an integrated digital network against itself, drew attention to how development of the ISDN meant

> the barriers between services are crumbling (voice, vision and data are indistinguishable in digital form; films made for the cinema may receive their first showing on TV or on video-cassette). The barriers between delivery mechanisms should also crumble. The screen and telephone are oblivious to the technology that lie behind them—as are their users! Thus a call to a mobile telephone in the field of a farm might come by a satellite from Hong Kong to a Mercury dish, through a BT line onto a cellular radio system.[93]

The DTI's report is interesting because it shows the flexibility in how standards can stand in for public policy. Standards have politics. For example, the adoption of a new internet protocol, IPv6, became a vehicle for furthering European integration and promoting economic growth in Asia.[94] The example of ISDN shows that standards' political flexibility can be pushed to contradictory limits. For European PTTs, ISDN stood in for a policy of public monopoly over telecommunications. But the UK's Department of Trade and Industry, rhetorically at least, used the ISDN to apply the principles of a free-market economy to innovation. Its report argued that, because ISDN separated the standard for digital transmission from the physical medium that carried the transmission, whether a satellite or optical fiber, ISDN promoted competition in the development of these media, so furthering innovation. For the DTI, the ISDN thus stood as a vehicle for applying a competitive marketplace approach to innovation policy, wherein the marketplace, rather than a national telecom operator, would determine which transmission technologies would succeed and which would fail.

After the government rejected a publicly funded fiber-optic "national grid," BT used that metaphor less, but kept promoting optical fiber as "highways" for the nation. BT announced the 1985 Nottingham–Sheffield fiber-optic link as a "high-capacity, long-distance highway," and BT described its fiber-optic network in 1988 as "a network of super highways of glass fibres for visual communications."[95] The highway metaphor also extended from the trunk network, which it had described with the waveguide, to the local network. By 1989, BT had installed 600,000 kilometers of optical fiber in the trunk network and so turned its attention to the local network and back to its failed plans to provide digital video services. BT started a fiber-to-the-home trial, or FttH, in 1989 in Bishop's Stortford, a market town approximately 30 miles north of London. One of the trial's novel features was providing broadcast TV and a video library, which meant that, alongside telephony and data, customers could also receive up to thirty TV channels and play TV shows and movies on demand.[96] This trial was advanced but not out of step with the ambitions of the time. Similar trials happened in Japan, Canada, France, the Netherlands, and the United States from the late 1970s to the early 1990s.[97]

This deployment of TV over optical fiber was a step toward BT's long-held ambition to incorporate TV and video services into its communication infrastructure, but not for long. In 1990, the Department of Trade and Industry began reviewing its competition policy for telecommunications. This included several topics, such as the BT/Mercury duopoly, but it also included the policy over whether BT and Mercury, as telecom operators, would be permitted to expand into cable TV supply. Concerned that the government would block this expansion, BT began slowing the deployment of optical fiber before any formal decision was announced.[98] This was reinforced in October 1990 as Peter Lilley, the secretary of state for trade and industry, announced that the government would be admitting more telecom operators, but gave no further details.[99] These details finally emerged in the government's 1991 white paper, *Competition and Choice: Telecommunications Policy for the 1990s.*[100] The white paper ended BT and Mercury's duopoly, allowing new companies to apply for licenses in local, trunk, and international telecom services. It also prevented BT from expanding into "entertainment services," meaning cable television, for seven to ten years, but permitted existing cable operators to immediately expand into telecommunications services. And so the DTI's plan for shaking up British telecommunications became clear. The government would not allow BT to expand into cable TV, but it would allow cable TV operators to expand into telecommunications.

This provoked various critiques through the early 1990s, which deployed all sorts of metaphors to advocate for a national fiber-to-the-home infrastructure. In response to the DTI, Alan Rudge, BT's managing director for development and procurement, called for public investment, saying that "telecommunications is the nervous system of society and we must nourish it properly."[101] Moreover, Rudge also made it clear that it would not be cost-effective for BT to roll out an all-fiber network, if it could not use that network to commercially offer entertainment services, including TV, over that network. Articles in the *Guardian* also lamented the government's decision, pointing out that, because there were no bars on foreign ownership of cable TV in the UK, this gave free rein to US-owned cable TV companies such as Pacific Telesis, US West, and United Artists, which had already bought many of the domestic UK cable operators, to expand into telecommunications as well. These articles called for fiber optic as both a "national grid" and an "information superhighway" for the UK, to little effect.[102]

It is telling, therefore, that BT transitioned the "highway" metaphor to a digital network that did enter service, albeit not in the form that BT had hoped. In 1998, BT's national ISDN service finally debuted under the brand name Highway, a broadband service that enabled customers to surf the internet and use their phone line simultaneously. Highway, providing simultaneous voice and data services to its customers, thus partly realized telecom engineers' original 1967 vision for a nationwide general-purpose digital infrastructure, albeit missing video services. Highway, however, also showed the influence of liberalization and business markets on BT since that original vision. Highway adverts told audiences to "find out how the business highway can help you work faster" and to "get on the BT highway."[103] A TV advert started with a car speeding down a road at night, before it suddenly stopped, its headlights illuminating a man sitting at his desk, using both his phone and computer simultaneously. The car reversed its journey until it returned to its starting point, illuminating the word "Drive."[104] In this advert, a physical, road-based journey was interrupted and reversed by a vision of a businessman making two simultaneous virtual journeys, telephonic and computerized. These adverts divorced the highway metaphor from optical fiber and, indeed, all transmission media altogether. Instead, the advert pitted metaphor directly against its real-life counterpart, with the message that the virtual, metaphorical highway was now superior to the material, physical highway, and that BT, in its new competitive condition, was the toll road operator.

This shows quite a different vision of the "highway" from Al Gore's popularization of the term "information superhighway" in the US in the 1990s, which Gore used to promote a national information infrastructure when he was vice president for Bill Clinton.[105] This usage reached its apogee at the 1994 Superhighway Summit at the University of California, Los Angeles. At this summit, Gore outlined his and Clinton's ambition for a national information infrastructure built on competition and private investment, while also ensuring open access and avoiding "a society of information 'Haves' and 'Have Nots.'" For supporters of Gore's highway, the metaphor connoted linearity, high speeds, and continuous flows.[106] Gore, however, also lamented the unexpected usages of the "information superhighway." A high-tech start-up had complained to him about the danger of ending up as "road kill on the information superhighway," while other companies had petitioned him to support their entry into the communications infrastructure market, concerned that they would end up "parked at the curb on the information superhighway."[107] These were not the only opponents of the "highway" metaphor. Contributors to the digital utopian magazine WIRED attached the "highway" metaphor as connoting the public ownership, monopoly, and technocracy of the federal highway system, and instead favored evolutionary metaphors that symbolized unpredictable growth and change.[108] BT, on the other hand, inverted the highway metaphor with its ISDN service. For BT Highway, the highway was not just a highway: it was BT's turnpike, and only in this new political economy of digitalization could such a metaphor, positioning private telecommunications above public roads, have appeared.

CONCLUSION

This chapter opened with the myth that Margaret Thatcher killed the UK's nascent national fiber-optic network and used a history of the Post Office and BT's vision of an integrated digital transmission environment to uncover the truth behind the myth. The myth has several inaccuracies. The first is that the 1991 decision was about optical fiber. It was not. Instead, it was about cable television and video entertainment services. For BT, optical fiber was a means to an end, and that end was broadcasting television signals and video services over a national cable network, whether coaxial or optical fiber. During the early 1960s, when the Post Office built a nationwide microwave network that transmitted telephony and television, its engineers and managers showed little interest in integrating television into their network. This changed in the late 1960s, when their understanding of digitalization and integration led them to the conclusion that they could build a network that could transmit video and data services in addition to voice telephony, and that doing so could not just preserve but also expand their monopoly.

This led to early interest in the integrated services digital network standard. This cooled when the Post Office's business customers demanded digital data networks sooner than digital telephone networks could integrate data services, and when transnational standardization efforts focused on integrating only telephony and data, and not video as well. The Post Office and BT remained interested in expanding into television as a way of realizing their original vision of a general-purpose digital infrastructure—this is something that the myth misses. *TechRadar* and Peter Cochrane's story about the 1991 decision to block BT fixates on the idea that this crippled Britain's internet infrastructure, but the decision had nothing to do with the internet and everything to do with television. This is further shown by the fact that in 2001, after the ten-year restriction on BT entering the TV market ended, BT began developing for cable TV. In 2006, BT finally entered the TV market, delivering TV via its customers' internet connections, realizing a level of digital integration that it had aspired to since it was the Post Office in the 1960s.

The second inaccuracy in the myth is how close the UK came to having this fiber-optic network. A national general-purpose digital network, which combined telecommunications and television, was only ever a dream on BT's part, and it had only ever been a dream since the late 1960s. BT, by its own admission, could have made the fiber-optic network cost-effective only if it could expand into TV provision. But time and time again, throughout the 1970s and 1980s, the Post Office and BT had lobbied both Labour and Conservative governments to let them expand into television transmission, an effort that had never, at any point, shown any sign of materializing. Indeed, if there was ever a moment where the national fiber-optic network came close to living, it would not be the 1991 decision about cable TV regulation. Instead, it would be in 1988, a moment of genuine conflict between government bodies on the public funding of a national fiber-optic grid. The Advisory Committee on Science and Technology supported the grid, while the Department of Trade and Industry's *Infrastructure for Tomorrow* report advised against it, using the ISDN standard to justify policies of infrastructural austerity and innovation through competition. The DTI, the senior government body, won, and the national fiber-optic grid never lived. The metaphors used throughout this history, highways and grids alike, reinforce the ephemerality of this national general-purpose digital network. These metaphors invoked the envisioned applications of waveguides and optical fibers to national communications, but they were only ever anticipatory. The metaphors that fueled the Post Office and BT during this period, like the other forms of long-range expectations such as the UK Trunk Task Force's modeling studies, reveal the extreme degree to which the Post Office and BT were inventing for the future, rather than for the present.

The third and final inaccuracy is that the 1991 decision was solely about competition. The 1991 decision is historically interesting because it shows an ideological quality missing from earlier decisions in the UK about the regulation of telecommunications. As these chapters explore, these decisions were usually pragmatic and often deferred to the Post Office or BT (as shown in the previous chapter on the regulation of equipment supply). For example, the decision to replace BT's monopoly with a duopoly sufficiently demonstrates that the Thatcher government's decision to privatize BT and liberalize its monopoly was not an ideological decision about turning British telecom services into a hypercompetitive free marketplace. As discussed in chapters 3, 6, and 7, liberalization and privatization were more about enabling BT and the City of London's financial sector to compete on international markets. The 1991 decision did open telecommunications to more competition, but by replacing the duopoly system, comprising majority British-ownership companies Mercury and BT, with a system that opened British telecommunications to foreign-owned cable companies. This was not just about competition but also about denationalization. As commentators both past and present have noted, the regulation of British industry under Margaret Thatcher increasingly showed that it was not the nationality of capital that mattered but attracting investment from international capital.[109]

The 1991 decision also shows a tragic irony in the intersections among digitalization, liberalization, and privatization. In the case of cable networks, it was possible to open telecom markets to cable operators only because digitalization meant that cable operators could integrate telecommunications services into their television networks. From the 1960s to the 1990s, the Post Office and BT hoped that digital integration meant they could expand their telephone network into a data and television network. But by the 1990s, when new cable television networks had appeared, digital integration also meant that these cable operators could expand into telecommunications. In the end, digital integration did not protect the Post Office and BT's monopoly, but instead rendered it more vulnerable to competition from alternative networks.

Integration, however, was not the only way that digitalization intersected with the liberalization and privatization of British telecommunications. The next three chapters use spatiality as an organizing motif to explore how the Post Office and BT's technopolitics of digitalization, nationalization, and privatization unfolded in different places and spaces. A recurring theme is the interplay between techno-nationalism and digital internationalism on these different levels, from the local to international. The next chapter starts with the local, looking at the Post Office and BT's research center in Martlesham Heath, Suffolk, one of the main sites where the Post Office and BT developed and tested waveguides and optical fiber.

III PLACES

5 MARTLESHAM HEATH: NOSTALGIA, FUTURITY, AND IT PARKS

Above the entrance to BT Labs in Martlesham Heath, Suffolk, is a plaque engraved "Research is the Door to Tomorrow." BT Labs is now the center of a science park, Adastral Park, but it was not always so. The plaque is a token from BT Labs' predecessor, the Post Office Research Station, in Dollis Hill, northwest London. The Queen formally opened BT Labs in 1975 as the Post Office Research Centre, after moving from Dollis Hill in North London, but Adastral Park is not Martlesham Heath's only distinctive feature. From 1975, an "instant village," built like "an unspoiled traditional village," was also constructed on the heath, in part to provide housing for Post Office research staff.[1] Martlesham Heath thus seems an anachronism, containing a future-facing plaque inherited from the past and a "traditional" village built in an instant. These temporal contradictions appear to emerge from several spatial changes: the relocation of the Post Office's Research Department, the construction of a new village, and the development of the science park, a quintessential "information age" development. In 1964, Martlesham Heath was not a place. Now, it is a place with its own sense of space and time.

How did this patch of scrubland in rural Suffolk turn into a science park and new village? Answering this question exposes how the relationships among digitalization, spatiality, and political economy have changed. The digital industries have been at the center of several novel spatial forms in the latter twentieth century, which have both shaped and been shaped by wider political economy. Silicon Valley looms large in this history, thriving on military patronage, ties to higher education, and a friendly entrepreneurial environment.[2] The US government, both state and federal, played an essential role in growing these digital districts, from Silicon Valley to Minnesota's "digital state" and Internet Alley outside Washington, DC.[3] These histories show that these spaces were never purely business endeavors, but were also shaped by political and urban history. Silicon Valley reinvented itself with Ronald Reagan's free-market politics, while highways and shopping malls shaped Internet Alley. These spaces also stand in for the nation. Silicon Valley's Mission Revival architecture, for example, ties it to a wider regional campaign of naturalizing settler colonialism in California.[4] This holds true not just for the digital industries, but all industries of national importance. French nuclear sites at Marcoule in the Gard region and Chinon in Touraine, for example, became regional spectacles that "brought the nation into the region," reconciling modernity and tradition in narratives that portrayed the French national nuclear industry as uplifting these regions.[5] But to return to digital industrial spaces, these histories show that they are more than simply stories about the economic successes or failures of industrial clustering. They also show that these places provide a greater understanding of how national politics and economies manifest themselves spatially and narratively. The history of Martlesham Heath thus offers a richer understanding of the relationships among national ownership, privatization, and digitalization.

This means looking at the history of Martlesham Heath and Adastral Park not just as an industrial space but also as a deliberately planned residential-industrial space. Recent history is full of examples of such spaces. Throughout the twentieth century, towns and cities across the Western world, from British garden cities to Italian "città di fondazione," were constructed as "techno-cities," which attempted to reconcile industrial advances with the "lost virtues of village life."[6] In the UK, "brave new towns" reconstructed urban spaces and dispersed the population from the 1940s to the 1960s.[7] These new towns have been called a "concretopia," hamstrung by utopian delusions of scientific planning and high modernist architecture, but it would be a mistake to see Britain's new towns as purely modernist projects. The new towns movement, and its intellectual ancestor, the garden city, were also shot through with nostalgia and conservative values. For example, the neighborhood unit concept of Clarence Perry, the influential US urban planner, projected a "seductive vision" of family-oriented villages built into cities.[8] Even those projects that were unabashedly modernist, such as Milton Keynes, the swansong of Britain's new towns movement, were dynamic and mutable.[9] Milton Keynes did not reaffirm scientific, welfare state urban planning but rather shows how the welfare state evolved and adapted to reactions against planning. In doing so, it shows how social democratic actors adapted to the market turn.

With the market turn came new approaches to urban planning and industrial space. The "enterprise zone" and the "new village" characterize the Thatcher era. Enterprise zones were the "purest policy expression" of the neoliberal city and began in Britain in 1981 in places like London Docklands, which, as the following two chapters discuss, also had a minor starring role in BT's international and commercial ambitions.[10] Enterprise zones suggest a neoliberal transformation of Britain's urban spaces, but "new villages" complicate this narrative. New villages, planned as small private-sector new towns that would take advantage of planning permissions deregulation, were planned in the 1980s. Tillingham Hall, a pioneer new village in Essex, became a "cause célèbre."[11] The new villages failed, however, because developers overestimated the support from Thatcher's governments in overriding stringent local planning practices. Martlesham Heath shows a successful "new village" almost ten years before Tillingham Hall, a new village that shaped and was shaped by the parallel development of a science park. So Martlesham Heath was neither a company town, like Lever's Port Sunlight or Cadbury's Bournville, nor a public-sector new town, like Harlow or Milton Keynes. Instead, it demonstrates a new settlement between public and private urban planning. Its history thus gives greater insight into the Post Office and BT's role in remaking Britain's residential and industrial spaces through the market turn.

As the quote on the BT Labs' plaque suggests, however, this is a question not just of space but also of time. Digital industrial sites and new towns, from Silicon Valley to Milton Keynes, are stereotypically future-facing spaces, but as Silicon Valley's Mission-style architecture, France's nuclear sites, and Clarence Perry's neighborhood units show, tradition and conservatism also mattered. Indeed, without historicity, neither organizations nor their places would exist. Organizations need historical discourses to support a lasting organizational culture, but this, in turn, requires maintaining and controlling material artifacts and sites, from paper records to brick-and-mortar buildings, that support these discourses.[12] Similarly, place-identity is constructed, in part, through retrograde temporal expressions, such as historically evocative architecture.[13] It would be easy to see sites like Adastral Park and Silicon Valley as placeless "spaces of flows," characteristic of a new flexible, networked, digital world.[14] This view misses, however, the centrality of place and historicity to making these sites. Without regional histories and identities, from the North Virginians' Civil War to the chateaux of the Torangeaux, there would be no Internet Alley in Tysons Corner, North Virginia, nor Chinon Nuclear Power Plant in Touraine—at least, not in the forms that they are now known. Exploring the role of historicity and place, from region to nation to organization, gives a way to see Adastral Park and Martlesham Heath as more than just instant, flexible spaces, but as places that are as much a part of BT's history as BT is part of theirs.

This chapter first explores the relocation of the Post Office's Research Department from Dollis Hill to Martlesham Heath, showing how central government used the Research Department as an instrument of spatial reordering. The second section analyses the construction of the research center at Martlesham Heath, showing how the new universities of 1960s Britain and the US postwar corporate research campuses shaped the new site. The third section moves on to the history of Martlesham Heath new village, showing how the village and the research center shaped each other, while also showing how the village's novelty and instantaneity referenced English "tradition." Finally, the chapter addresses Martlesham Heath's recent history by exploring the creation of Adastral Park, which showcased a new transnational spatial strategy for research in the wake of BT's privatization, but which was also rooted in novel representations of the place and history of Martlesham Heath.

RELOCATION AND DISPERSAL

By the late 1950s, the Post Office wanted to relocate its research department from Dollis Hill, northwest London, for two reasons. First, the research station was over capacity. By 1958, 1,200 staff worked on a site built for eight hundred, and studies projected that, by 1974, numbers would increase to almost two thousand.[15] Second, London's suburbanization had brought electrical and vibratory interference.[16] The Post Office had built Dollis Hill in rural environs in 1919, and although it sat close to the Metropolitan Railway, it was distant enough to avoid interference from passing trains.[17] By the 1950s, however, the surrounding population had reached around 316,000, from around 140,000 in 1906, and from the mid-1930s, the London Underground's Bakerloo line began to run more services through the area, taking over for the Metropolitan line's congested Stanmore branch.[18] As figure 5.1

FIGURE 5.1

Dollis Hill in 1914 and 1936. The box shows, in the top map, the planned site for the research station in 1914, the same year its purchase was originally authorized, and, in the bottom map, its actual site, surrounded by suburban London, in 1936. Maps from the Ordnance Survey 25-inch County Series for England and Wales reproduced with the permission of the National Library of Scotland.

shows, between 1914 and 1936, Dollis Hill turned from rural farmland into suburban London.[19] Gordon Radley, the Post Office's director-general, decided to relocate the research station in 1958 and set three requirements for a new location. First, it should be close enough to maintain good contact with the Engineering Department, especially development, in London. Second, it should be close to towns with residential facilities and good day schools. Third, there should be a technical college nearby for further education for staff.[20] Each of these requirements—the boundaries between research and development, the relationship with housing developments, and the links between research and higher education—shaped the place that Martlesham Heath would become.

The Post Office general directorate's first choice was the Harlow new town.[21] Harlow did not pan out, but its selection is suggestive of the Post Office's search philosophy. Harlow was one of the first wave of postwar new towns, which became vehicles for visions of a modern, reconstructed Britain. As noted in the *Architects' Journal*, Harlow was one of the foremost examples of novel town design concepts, including Clarence Perry's "neighborhood unit," where residential neighborhood units were isolated from arterial traffic routes to engender community spirit.[22] Other electronics companies had relocated to Harlow, including A. C. Cossor, later acquired in 1961 by the US defense contractor Raytheon in 1958, and Standard Telecommunication Laboratories, the UK-based research center for STC and its parent company, International Telephone & Telegraph (ITT), in 1959. Harlow clearly held some appeal for electronics R&D, so it may have been on the Post Office's radar as a potential cluster of expertise. Dollis Hill staff rejected Harlow, however, claiming that it lacked adequate housing and education and that the Post Office had not explored alternatives.[23] The Post Office thus formed a joint management-staff relocation working party in 1962. While Harlow did not meet staff's expectations, it does show that, from the beginning, the relocation and construction of a new research center were, for the Post Office, bound up with the colocation of corporate research, as well as contemporaneous visions for a "new" Britain.

The search for a new site also became embroiled in central government initiatives to develop employment opportunities around Britain. In 1963, the Flemming report, "Dispersal of Government Work," identified Civil Service departments suitable for relocation outside London.[24] The report identified the Post Office research department as a prime candidate, forcing the Post Office to commit to a site outside London. A relocation memorandum produced after the report noted that dispersal's goal of relieving regional unemployment "was a secondary consideration which has recently assumed perhaps more importance."[25] One potential factor that might have motivated dispersal was the changing ethnic makeup of London and other British cities. The 1962 Commonwealth Immigrants Act, intended to reduce immigration, had led to an influx of women and child migrants, who were exempt from immigration controls if they were accompanying a family member who already resided in the UK. This led to media hysteria about ghettos, inspired by the Jim Crow South and South African apartheid, and various racialized dispersal policies, such as bussing immigrant school children and Birmingham City Council's illegal social housing dispersal policy for Black Caribbeans.[26] But the dispersal of middle-class government work also evokes the image of "white flight," and both the influx of people of color and efflux of white middle-class workers from cities influenced the postwar rise of rural and suburban industrial sites for skilled workers in the US.[27] Dispersal was not always just about employment, but also about race and class. In Dollis Hill's case, the motivations—inadequate space, electrical interference, regional unemployment—appear to have been more pragmatic, but as these other histories show, race and class cannot be so easily dismissed as factors in dispersal.

By July 1963, the Joint Working Party had finished its shortlist. Hastings, Sussex, was at the top, using the same criteria, as established in 1958, of towns with a suitable housing supply, a good number of schools, and a technical college for further education. Whereas previously, however, the Post Office had required the new site to remain close to engineering and development in London, this condition was relaxed to include regions up to two hours' commute from London, bringing the move in line with dispersal.[28] Hastings Council proved very welcoming, informally confirming to the Post Office that they would cancel a planned aerodrome development to provide space for the station and prevent interference from aircraft—although, in a display of faith in the British aerospace industry, one council member did ask whether there would still be space for a small landing pad for vertical-takeoff aircraft.[29]

A leak, however, from the Ministry of Public Buildings and Works (MPBW), that it planned to move its offices to Hastings, stalled the Post Office relocation. This leak irritated Post Office senior management, as the MPBW had been aware of the Post Office's plans, but the Post Office was still keen to secure Hastings and avoid losing face with the staff.[30] Management attitudes changed once the MPBW officially announced its move in January 1964 and was, in the words of Postmaster General Reginald Bevins, "roasted" in the House of Commons by MPs from Scotland, the North-East, and the South-West.[31] Dispersal, a program designed to alleviate unemployment, had not been created for the affluent South-East, so relocating to Hastings appeared politically untenable. This was particularly worrying for the Post Office as other shortlisted sites were also losing viability. Sites in Christchurch and Poole, Dorset, were on difficult terrain, while a potential site near Ipswich, Suffolk, was becoming "restive" as the landowners, Bradford Property Trust, were keen to start a residential development, the soon-to-be Martlesham Heath new village. A site in Folkestone looked uncertain after the somewhat premature announcement that Folkestone would become the British end of the Channel Tunnel, which would open thirty years later.[32]

MPs from the North and Scotland also courted the Post Office. Jeremy Bray, MP for Middlesbrough West, advocated his constituency to the Post Office. A delegation of Scottish MPs made an impassioned case for Scotland, pointing out that the Department of Scientific and Industrial Research had a site in East Kilbride, Ferranti had a site in Edinburgh, and National Cash Register had a site in Dundee.[33] Tam Dalyell, MP for West Lothian, even formulated research agendas, suggesting that Scotland would be ideal for researching masers. The Post Office told both Bray and the Scottish delegation that the research station needed to be closer to the Engineering Department and did not reveal that staff would not move northward, nor that the Post Office did not believe it could recruit new staff in the "less attractive" North-East.[34]

Instead, the Joint Working Party turned its attention to Ipswich, third on the list, after discarding Christchurch because of the site's terrain issues and because there were already two government research establishments there, the Military Engineering Experimental Establishment and the Signals Research and Development Establishment.[35] Again, dispersal overrode any other considerations, such as the benefits of being near other government R&D laboratories. Ipswich was suitable because, four miles east, there was a promising site on a sizeable former airfield on Martlesham Heath, an active RAF and USAF airfield during World War II and an experimental aviation site before and after the war. Furthermore, a South-East development study had earmarked Ipswich for major expansion, thus alleviating staff concerns about housing availability.[36] Finally, the new University of Essex, established in nearby Colchester in 1963 as part of seven "plateglass" new universities—East Anglia, Essex, Kent, Lancaster, Sussex, Warwick, and York—fulfilled far and above the Post Office's hopes to forge links with a local technical college.[37] The Post Office secured staff approval in June 1964 and announced the move to the House of Commons in July.[38] Two of these motivations, the development of the Ipswich area and the new University of Essex, were especially influential on the research station's subsequent development.

INDUSTRIAL VERSAILLES MEETS THE NEW UNIVERSITIES

Two different types of postwar research site—the UK's new universities and US corporate research laboratories—influenced the new research center at Martlesham Heath, particularly its rural aesthetic on the "heath." In planning the new site, the Post Office forged explicit links with the new University of Essex while also making implicit references to the new universities program. Press releases announced the Post Office's hopes for a "university character," which was "of a clean and quiet nature and in appearance might resemble a university area with mainly low buildings set in landscaped areas."[39] These hopes echoed the aesthetic of Britain's new "plateglass universities." Referring to their common aesthetic of plateglass facades held in steel or concrete frames, these plateglass universities were established in the 1960s as a defense of modernism and social democracy, another welfare state public project in step with the social architecture of public housing and new towns like Harlow.[40] Bearing a striking similarity to the Post Office's vision for Martlesham Heath, several early new universities—Sussex, Warwick, Kent, and York—were built on landscaped parklike campuses, composed of low-profile buildings of metal, glass, and concrete. The government launched the universities with fervent publicity that encouraged the nation to appreciate the "social importance" of these new institutions.[41] The Post Office's invocations of "university character" and aesthetic visions of low-profile buildings on a landscaped campus suggest that it aspired to imitate these new, socially important institutions. This echoes the Post Office's earlier selection of Harlow, another example of its desire to align the new research center with the state's modernist, social democratic projects for a "new" Britain.

The Post Office forged strong ties with the University of Essex, the closest new university to Martlesham Heath. The University of Essex took on a more "integrated urban" form compared to its predecessors, and was perhaps the most high-profile of all the new universities—"none of Britain's postwar universities was launched with such enthusiasm."[42] The Post Office was quick to forge ties after the university opened in 1964. In 1967, the university established a chair and a lectureship in telecommunications systems with a grant from the Post Office.[43] Collaborative MScs and PhDs and research grants followed, including £30,000 in 1971 to research video telephony.[44] The Research Department set up an MSc in telecommunications, in which junior engineers enrolled and senior engineers worked as lecturers. The University of Essex, compared to the other plateglass universities, later took on a more ambiguous public presence, as it became a central and controversial site for the British wing of the 1968 student protests, but the Post Office and, later BT, nevertheless continued partnering with the university into the twenty-first century.[45]

The Post Office also took cues from US corporate laboratory design. In 1964, representatives from the Post Office and the MPBW, which would provide architectural services for the new research center, visited laboratories across the US, including Bell Labs, the Radio Corporation of America's laboratory, Hughes Research Lab, Fairchild Semiconductor's R&D labs, and the Stanford Research Institute.[46] Two features that the British delegation found most noteworthy, artificial lighting and air conditioning, later caused friction between management and staff at the new research center. The delegation heard that air conditioning was essential for controlling atmospheric conditions for delicate electronics work, while artificial lighting allowed a more compact building design, allowing internal rooms without windows. The Post Office report on the trip described this as "Americans have no objection to working in rooms without natural light."[47] The other significant US feature that the Post Office would emulate was flexible partitioning, which researchers could use to reconfigure rooms. Of all the visits, the most influential were to the laboratories designed for Bell Labs and IBM by the Finnish neo-futurist architect Eero Saarinen. Saarinen's research centers for GM (called a "Versailles of Industry" by *Life* magazine), IBM, and Bell also emulated the isolated campuses of US postwar universities.[48] In doing so, these corporate laboratories, as with postwar US university campuses, embodied a linear model of research that saw vast centers of basic research as generating scientific knowledge, which would later fuel technological development. This model of research influenced postwar universities and US research campuses, which in turn influenced Martlesham Heath. It would also influence the Post Office's reorganization of R&D, which accompanied the research department's move to Martlesham.

The plans and construction of the Martlesham Heath research center implemented these influences from the new universities and US corporate research campuses. The Post Office established the main design features in 1967 with the MPBW, which provided architectural and design services to meet the Post Office's schedule of requirements.[49] The plans initially featured a lagoon, partly as a landscape feature and partly as a resource for fire engines, but the final plan dropped the lagoon for costs.[50] The lagoon plan evoked both the landscaped campuses of the plateglass universities and US corporate campuses. York and Essex had pools and lakes, and all the new universities' campuses were landscaped to varying extents to create "park" atmospheres, while Saarinen's Bell Labs site in Holmdel, New Jersey, also had a large lagoon.[51] Departing from the new university aesthetic, and perhaps acknowledging the Cold War reasons for dispersing government work, which were more visible in the US, the Post Office and MPBW briefly discussed, and then discarded, plans to build a fallout shelter.[52] Unfortunately, the plans did not explain the aesthetic decisions taken in the research buildings' architectural design, leaving the impression that form followed function. Like US corporate labs, the buildings used extensive air conditioning and flexible partitioning. Staff, however, resisted plans for artificial lighting, leading to more windows and natural light in exterior rooms. Internally, the new research center mainly re-created US corporate labs, while the landscaped, low-profile appearance on the outside evoked the new university campus.

The research center's eventual form was three main buildings: a seven-floor lab block, a three-floor administration building, and a single-story building to accommodate mechanical engineering, a drawing office, a workshop, and storage. Two towers, for radio and water, were also constructed.[53] The towers, which served as the primary lift shafts, sat on opposite sides of the main lab block to encourage circulation through the building. This goal echoed the idea that movable partitions would encourage "flexibility" in research.[54] A separate acoustics complex contained two anechoic chambers, a reverberant room, and a microphone calibration room, and achieved acoustic isolation with double-walled construction that fitted the inner rooms on acoustic mountings.[55] In 1968, staff arrived and construction began. A small research team worked on waveguides out of temporary huts while the contractor, Mitchell Construction, began work. The research center finally opened in 1975 after Mitchell Construction's bankruptcy caused delays, but construction resumed after Tarmac bought out Mitchell.

Building a new research center also shaped the organization of R&D in the Post Office. Before the research center's move to Martlesham Heath and before the Post Office's corporatization in 1969, research and development were separate branches within the Engineering Department. Strictly speaking, they were on the same tier of the organizational hierarchy, but research appears to have had a higher status. In 1946, for example, the Post Office set up a combined Research and Development Subcommittee chaired by the controller of research.[56] During the relocation to Martlesham, after the Post Office became a corporation, the Post Office dissolved the Engineering Department and set up a Development Division, in which the Research Department and Development Departments became subunits. This growing emphasis on development appears to emulate trends from across the Atlantic. "Research and development" gained prominence in the US after World War II, popularized by the US Office of Scientific Research and Development, founded in 1941 for the war effort. The term followed the further integration of research into industrial "development."[57] "Development" emerged as a keyword alongside research because, from the midcentury, corporations began to differentiate internally between research and development and gained prestige by combining research and development to show off gross R&D expenditure.[58] As the US corporate research campuses of the postwar era suggest, this was not simply a bureaucratic exercise but also a spatial one. The US "industrial Versailles" illustrate how GE, AT&T, and IBM literally built a linear model not of research *and* development but of research *for* development.[59]

The Post Office's growing emphasis on "development," placing the research department and, by extension, the research center, within a development division, was not simply a matter of emulating the US model, however. This reorganization collected a series of spatially dispersed departments. Previously, Research Branch and Development Branch were both in London. By the mid-1970s, various development division departments were strewn across London and the East Anglia Complex, composed of Martlesham Heath and a long-range planning unit in Cambridge.[60] Emphasizing the "D" of R&D was not a matter of organization charts, but rather a way to give these departments a group identity across space. The spatial dislocation of Post Office research had created the need and opportunity to form new organizational identities for research and development. This was clear in how Martlesham Heath aspired to appear like a new university, how it imitated US corporate laboratory design, and how the Post Office emphasized development. That said, these were not the only influences on Martlesham Heath's place-identity. In 1975, a new village began construction right next to the Post Office Research Centre.

THE INSTANT TRADITIONAL VILLAGE

The Research Department's relocation to Martlesham Heath created a local need for housing. In 1964, a government study of the South-East identified Ipswich as a potential major development area. The owners of the Martlesham Heath site, Bradford Property Trust, which had sold a section of their land to the Post Office for the new research center, also planned a residential development and, in 1965, commissioned development plans from the architectural firm Clifford Culpin.[61] The proposed development was for ten thousand people, and the relocation of the Research Department took a prominent place in the report. The proposed town encircled the research site, which "would allow the GPO to be related to the heart of the town."[62] The report also pointed to the Post Office's emphasis on university character as a "splendid basis on which to build a new community" and suggested leaving room to develop light industries in keeping with the "university background."[63] The role of new infrastructure in developing the Suffolk region also featured. The proposed development's layout accommodated future transport infrastructure, including a monorail station in Martlesham Heath.[64] The developers also envisioned an exclusive, communal atmosphere, seeing an opportunity to use "the best forms of social and architectural planning in order to relate the needs of the individuals to those of the community." They thus proposed a community trust to safeguard the development.[65] Landscaping, as with the research center, would also create and preserve this unique atmosphere. Tree planting and earth mounding would "maintain and enhance existing views" and would also "screen certain development" around the village.[66] A contradiction thus emerged, wherein the research center and proposed industrial development were central to the plan, yet the community vision was best served by screening those spaces.

Central government, however, did not pursue the South-East expansion proposals for Ipswich, delaying the new village. East Suffolk County Council then had to choose between approving the de novo Martlesham Heath development or expanding Kesgrave, an existing village slightly farther east.[67] These delays led to the Research Department's Move Committee lobbying the County Planning Officer to accelerate the decision-making process and choose Martlesham Heath over Kesgrave.[68] In December 1972, the county council finally approved the new village, and construction started in 1974. But between the initial development plan and the construction of Martlesham Heath, the vision had changed substantially. The new plan embodied design ideals that had not appeared in the 1964 plan. Martlesham Heath would be, on the one hand, a "traditional village," and on the other hand, an "instant village" or a "new village."[69] The new aim was for a "twentieth century village" that reflected the region's design traditions by using Suffolk vernacular stylings, while also deliberately changing scales, street patterns, and rooflines to give the appearance of having developed organically over time, as "is so evident in many villages in the country."[70]

Martlesham Heath's lead architect, Christopher Parker, described the village as a "revolt against convention," specifically against the highly planned postwar housing of new towns that he argued had failed through an "insistence on control and careful avoidance of any design function."[71] Parker chose the village concept instead, using an "incoherent" architecture that would emulate real villages. The "instant village" concept would put "as much village character into the design as possible" while also using contemporary building techniques to accelerate construction.[72] Parker also achieved the traditional village concept through broader planning and landscaping. There was a central village green with a cluster of commercial units for a village store and pub.[73] There were still some continuities with the earlier plans. Parker kept the neighborhood unit concept, organizing the new village into various "hamlets" built off the village's arterial loop road. Despite this conceptual lineage from US interwar planning and postwar new towns like Harlow, which also used the neighborhood unit, these hamlets also served the village's chaotic aesthetic choices. Each hamlet had distinctive and varied design concepts while still adhering to the traditional English aesthetic and Suffolk vernacular style.[74]

The developers also achieved this aesthetic by limiting technology's intrusion into the traditional village aesthetic. Bidwells, the site developer, used earth mounding and tree planting to exclude road noise and screen the research center, which magazine features on Martlesham Heath described as a "huge concrete toadstool of a research centre" giving a "totally unbalanced impact" to the new village.[75] Through the community association set up by Bradford Property Trust, residents had to sign covenants banning them from putting up TV aerials and from parking caravans in the village. Villagers instead received TV reception via an underground cable carrying a television signal from an aerial atop the research center's radio tower. A block of flats in the village center had "their parking courts discreetly positioned to avoid the visual intrusion of the car."[76] Martlesham Heath's development thus captures one of the essential characteristics of the twentieth century's "techno-cities," a "techno-nostalgia" that combined high technology and preindustrial Eden through practices such as concealed television infrastructure and vernacular styling.[77]

These design choices also reflect postmodernist trends in architecture and urban planning. The emphasis on vernacular and regional style came as part of a postmodern neo-Vernacular movement in 1970s British architecture, which was "the style to fall back on when there were no other clear directions" and a "dreamscape" for an old English palette.[78] In this sense, Martlesham Heath resembled other developments, such as Woodham Ferrers, Essex, and Poundbury, Dorset, that aimed to capture the traditional English aesthetic. These developments also earned critics. Woodham Ferrers was called a "pastiche," and Poundbury's revivalist aesthetic, designed by the postmodern architect Leon Krier and supported by Prince Charles, was called both "radical" and a "cottagey slum."[79] These attempts to engineer history and tradition in brand-new developments reflect the notion that postmodernist styles, such as vernacular housing, responded to accelerating times by turning to the stability of the past.[80] A review of Martlesham Heath in the *Royal Institute of British Architects' Journal* articulated that "the perceived stability of the vernacular idiom in housing design is an understandable reaction to a rapidly changing world."[81]

Martlesham Heath's contradictory combination of "incoherence" and "stability" captured the essential tensions of postmodernism, well-put by a 1983 *Chartered Surveyor Weekly* feature on Martlesham Heath titled "Controlled Chaos Proves a Winner."[82] Herein lies a key point about the new village. Superficially, the village represented incoherence, but this aesthetic required various strategies of control, some of which drew on the research center, to give the appearance of organic evolution and natural integration into the Suffolk region. These strategies were organizational, forming a householders' association; spatial, deploying a neighborhood units/hamlets concept; architectural, in the incoherent yet local vernacular style; environmental, in the landscaped screening of the research center; and technological, using the research center for TV reception. As these last two strategies show, this aesthetic was not independent of the Post Office Research Centre. Martlesham Heath's place-identity was a dialogue between the new village and the research center. This dialogue would continue with the research center's transformation into a science park after BT's privatization.

ADASTRAL PARK: HISTORY IN THE MAKING

Liberalization and privatization affected both the research center and the new village, bringing new ways to historicize Martlesham Heath's place-identity. Liberalization and privatization meant that the research center, renamed BT Research Laboratory in 1981, became more oriented toward business markets and commercializing research. In 1981, the financial advisers Lazard Brothers proposed a subsidiary called Martlesham Enterprises to BT's board.[83] Martlesham Enterprises, founded in 1982, sponsored and secured financing for research spin-off projects that were peripheral or irrelevant to the telephone network but still commercially viable.[84] The Thatcher government's emphasis on innovation, small enterprise, and the "sunrise" IT industries particularly influenced the proposal, which was a common strategy in the 1980s and 1990s. BP and ICI also founded similar subsidiaries, BP Ventures and Marlborough Technical Development.[85] Another 1980s industrial strategy, interlinked with the Thatcher government's IT boosterism, was the "science park," a special-purpose cluster of industrial and academic research centers. Indeed, science parks' early 1980s popularity was significant enough that the *Financial Times* ran a special section on science parks in 1983 and referred to Silicon Valley as a prime example, emphasizing its relationship with Stanford University's research campus.[86] Again, the links between postwar universities and research sites inspired new spatial forms, but it was not until the 1990s that BT drew on the science park concept.

Throughout the 1980s and early 1990s, Martlesham Heath became a site for collaborative commercialization of research. BT ran open days, called Innovations at Martlesham, showcasing research to financiers from the City of London to garner investors. BT Labs also began to offer consultancy services in which research staff would consult for other companies on research and development. BT embarked on various collaborative ventures and contractor arrangements with companies from across the world, including DuPont, Corning, Mahindra, and AT&T, all of which brought work to Martlesham

Heath. BT started a collaborative venture with DuPont, BT&D, in 1986 to manufacture optoelectronic components for fiber-optic communications.[87] BT also set up a joint venture, Tech Mahindra, in 1986 with the Indian conglomerate Mahindra & Mahindra to provide technology outsourcing, which was colocated at BT Labs.[88] In 1994, BT acquired 20 percent of MCI, the US long-distance telecom provider, for $4.3 billion, and also bought the Minnesota-based Control Data Systems for its global services organization.[89] As part of this global services arm, BT formed a $10 billion joint venture, Concert, with AT&T in 1998 to provide network management services.[90] In 2000, BT also began a collaborative research partnership with Corning, the US glass manufacturer, which acquired BT's photonics lab at Martlesham Heath as part of the deal.[91]

In 1999, BT distilled this spatial colocation of its partners in and around Martlesham Heath into Adastral Park, a new science park. Adastral Park was part of BT's broader ambition to turn the site into a high-tech hub.[92] Adastral housed BT's technology and research partnerships, mentioned above, but also housed subsidiaries and spin-offs, such as Ignite, an e-business and communications solutions subsidiary; Napoleon, a joint venture with the private equity firm 3i to provide network management software; and Quip!, a web-based international phone call provider.[93] In 2000, BT set up a technology incubator, Brightstar, that took minority stakes in companies and provided advice, management services, and on-site accommodation.[94] BT also set up the East Anglia High Tech Corridor in partnership with Vision Park, Cambridge, where BT based a subsidiary, Internet Designers, which provided BT with internet, IP, and multimedia support services.[95] At present, Adastral Park houses many companies, including 3M, Cisco, Intel, and Huawei.[96] Huawei's partnership with BT has been controversial. In 2013, the British Parliament's Intelligence and Security Committee criticized BT's decision to award critical infrastructure contracts to Huawei, which the government saw as a security risk given its rumored associations with the Chinese government. Since then, the UK has announced a ban on new Huawei installations in its 5G networks, effective from September 2021, while the Australian government has excluded Huawei from involvement in Australia's National Broadband Network, and the US government has banned it from bidding for government contracts.[97]

Despite these partnerships and ambitions, Martlesham Heath did not become the UK's Silicon Valley. Neither did Silicon Glen in Scotland, nor

Silicon Fen in Cambridgeshire, with which BT Labs forged ties along the A14 road's East Anglia High Tech Corridor, from Ipswich to Cambridge, perhaps attempting to emulate Boston's Route 128 or North Virginia's Internet Alley. If any site has a claim to that in the UK, it is perhaps Silicon Roundabout, a cluster of digital industries in East London that started growing around Old Street Roundabout in the late 2000s. But that timing reveals the fallacy of comparing Martlesham Heath and Adastral Park to Silicon Valley. Martlesham Heath was never planned to be like Silicon Valley. Silicon Valley was no accident of the market, but an intentional "city of knowledge." It was funded by the US Cold War defense complex, which invested in economic development around a powerful university, Stanford, and was maintained by private-sector venture capital that replaced the public sector after the Cold War ended.[98]

Martlesham Heath, on the other hand, was not intended as a center of economic development. Much more modest were the ambitions of engineers, bureaucrats, and town planners to find a place for one research center and build a village for both engineers and locals. It never had the public-sector investment of Silicon Valley, and the partnership with the University of Essex was nowhere near the same scale of intervention as Stanford University's in the Bay Area. Even after privatization, when BT began to consciously emulate Silicon Valley by building a science park, setting up incubators, and attracting high-tech partners and venture capital, the effort was always that of just one firm. The creation of Silicon Valley shows how the spatial intersections of digital industries and capital changed in the US across and after the Cold War. That of Martlesham Heath, on the other hand, shows that the UK was never especially interested in these spatial intersections until after the public sector retreated from the digital industries. Instead, the UK's preferred approach was industrial, rather than spatial, creating national champions like ICL, through forced mergers.[99] The irony therefore is that Britain's telecom business tried to make its own Silicon Valley only after the window of opportunity provided by public-sector investment and intervention had ended.

This fact has not stopped a temporal process that tries to extend Adastral Park backward in time, back to the Research Department's relocation and before. The process has historicized Martlesham Heath as an innovative place and collided the place-identities of the research center and the new village. BT often describes Adastral Park as a "science campus," echoing the "university atmosphere" of the original Post Office Research Centre.[100] In 1967, a

Post Office report on relocation titled "Martlesham Heath: Home of Experimental Units" turned the heath's prior history as an experimental aviation unit into a further cultural justification for the move.[101] In January 1917, the Experimental Flying Section of the Royal Flying Corps' Central Flying School moved to Martlesham Heath, and the site remained an experimental aviation site until World War II, at which point the RAF used it as a forward base until 1943, when the RAF loaned it to the USAF. After World War II, it housed the Bomb Ballistic and Blind Landings Unit until 1961, when it was put onto care and maintenance status, finally closing in 1964, and the land reverted to the Bradford Property Trust. BT reiterated this experimental history in 1999 when it created Adastral Park, celebrating "30 years of BT research" at Martlesham Heath to inaugurate the creation of Adastral Park, but positioning this celebration within a longer lineage of research. An internal BT magazine described how RAF pilots had flown "research missions" from Martlesham Heath, narrating that "in those days, pilots like Sir Douglas Bader could be seen testing the latest Spitfires to destruction."[102] The relocation of the Post Office, readers were told, meant that Martlesham Heath "once again became a focus for leading-edge technologies, with Spitfires being replaced by circular waveguides and optical fibres." Tellingly, this publicity celebrated the establishment of Adastral Park as "history in the making."

This World War II heritage has become prominent in the historicization of Martlesham Heath. The thirtieth anniversary celebrations described the initial move to Martlesham Heath as an "expeditionary force" and played up Martlesham Heath's role in allied bombing raids.[103] The name "Adastral Park" is a deliberate reference to the RAF, referencing the force's motto "Per Ardua Ad Astra," which translates as "through adversity to the stars." Further invocations of World War II heritage have since appeared. Adastral Park now also houses the Tommy Flowers Institute, referencing Flowers's and the Research Department's role in World War II codebreaking at Bletchley Park.[104] The institute aims to foster collaboration between the ICT industry and academia in the UK, an ironic goal given the fractious relationships Flowers, a working-class engineer with no higher education, had with his academic Bletchley Park colleagues Alan Turing and Max Newman. The Tommy Flowers Institute, with its emphasis on academia, is thus a highly selective invocation of the past.

Martlesham Heath's historicization of innovation has continued to the present, converging with the new village's place-identity. Martlesham Heath village celebrated 2017 as MH100, an anniversary of one hundred years of innovation on the heath, sponsored by, among others, BT.[105] MH100 was celebrated through a July weekend fete on the village green, next to the Douglas Bader pub. Attractions included full-size Hurricane and Spitfire aircraft; vintage military vehicles; historical reenactments, including a Winston Churchill actor and impersonator; representatives from the RAF and USAF; World War II songs and jive dancers; and a Battle of Britain memorial flyby.[106] The event's website summarizes that "the story of Martlesham Heath is one of innovation, research and development, initially for aviation and more recently for telecommunications and IT. The Martlesham Heath 'new village' itself was an innovative approach to building new housing."[107] Reflecting the historicist turn in British community life, MH100 shows how the special occasions of everyday life, such as fetes, can be potent and selective enforcers of the past.[108]

The telephone business's research activity, which itself drew on a prior history of aviation research to cement its own innovative research identity, has thus created Martlesham Heath as a "place," a crystallization of time and space. The construction of place has required the construction of history, and that construction extends to this book, which began as a project, along with two others, in a joint Science Museum/BT Archives research proposal to investigate the history of the Post Office and BT's R&D. This proposal took a spatial approach to R&D, conceiving it as places, first Dollis Hill, then Martlesham Heath. As this chapter and book shows, however, R&D was a blurry and dispersed activity, including research at Dollis Hill and Martlesham Heath and long-range planning in London and Cambridge. The plaque above Dollis Hill and Martlesham Heath's entrances reads "Research is the Door to Tomorrow," but research at the Post Office and BT has, in many ways, also served as a door to a yesterday of national ownership.

CONCLUSION

Martlesham Heath's spatial transformations continue. In 2008 and 2009, BT submitted planning applications to build two thousand homes southeast of Adastral Park.[109] Suffolk Council rejected these plans and a group, No Adastral New Town, with ties to the Martlesham Heath householders' association, fought the development.[110] In 2018, however, East Suffolk District Council gave outline planning permission for the £300 million development, called

Brightwell Lakes.[111] This furor over "Adastral New Town" stands in stark contrast to the 1960s and 1970s, when enthusiasm for newness, new towns, new universities, and new villages characterized central government's social democratic reshaping of Britain, the Post Office's move to Martlesham Heath, and private sector development of the heath. The new village's private-sector origin shows that it was not just the social democratic state that had forward-facing visions for Britain, yet the influence of central government's reordering of Britain cannot be overlooked. The new universities and new towns both left their mark on where and how the Post Office relocated research away from Dollis Hill, to the extent that the arrival of a few researchers working out of huts on Martlesham Heath in 1968 must be seen as another example of the spatial dimension of social democracy.

But, since then, it has been nostalgia and not just novelty that shaped the formation of Martlesham Heath's place-identity. The new village concept departed from futuristic, new town visions of monorails and instead reacted against them, using "controlled chaos" to invoke local English tradition and heritage. The new village brought the past into Martlesham Heath, and BT wove that nostalgia into its organizational fabric. As the two sites met, embracing and obscuring each other, and as the Post Office Research Centre became BT Labs and then Adastral Park, the past took on an ever-greater role in Martlesham Heath's organizational identity and place-identity. These histories of research sites and places cannot be separated. Just as one cannot tell the history of the Physikalisch-Technische Reichsanstalt, the German metrological laboratory, without the history of urbanizing early twentieth-century Charlottenburg, nor the history of Britain's Jodrell Bank radio telescope without the history of suburbanizing postwar Manchester, so the history of Adastral Park cannot be told apart from Martlesham Heath.[112] This place is as much a part of BT as BT is a part of this place.

Martlesham Heath's new village and research center departed from the social democratic visions of British space associated with new towns and replaced that with a mix of private-sector nationalism and transnationalism. The private sector built Martlesham Heath, a techno-village that built on the twentieth century's longer tradition of techno-cities, mixtures of rural nostalgia and industrial futurism. Martlesham Heath scaled down the techno-city into a "traditional English village," still defined by the research lab next door, and offered an alternative of postmodern national and regional vernacular architectural stylings to a public that had allegedly tired of "welfare state modernism."[113] The creation of Martlesham Enterprises and a science park aligned with the Thatcher government's techno-nationalist focus on the sunrise digital industries, but there was hardly much direct support from the Thatcher governments. Instead, privatization empowered BT to develop transnational partnerships that focused on network technologies and services, from AT&T in the United States to Mahindra in India, all the while deploying national narratives about the RAF and World War II to explain the raison d'être for research on the heath. In short, the Thatcher governments had little to do with either Adastral Park or Martlesham Heath new village. Martlesham Heath shows how an increasingly privatized political economy shaped both new settlements and digital industries, and that the denationalization of corporate R&D rested on the nationalization of historical narratives.

These changes thus show the importance of place to understanding how corporate R&D has changed. The corporate research laboratory appeared around the start of the twentieth century as an expression of the vertical integration of research. After World War II, new types of R&D campuses appeared, from Eero Saarinen's Bell Labs to the Post Office's Research Centre, which materialized the linear model of research.[114] But this model no longer characterizes Adastral Park, which replaced the corporate laboratory with an R&D strategy that was spatially bipolar. In one instance, collaborative ventures took place on a "science campus" in Suffolk, and yet in another, they virtually span the globe, managing and managed through digital networks. Innovation in this model is not about the linear flow of basic research to technological development. Instead, it is about horizontal, transnational, corporate collaboration on advanced development and commercial products. Adastral Park thus shows British telecommunications' denationalization through venture capital and corporate collaborations. The next chapter explores these partnerships and Britain's digital denationalization further by looking at the Post Office and BT's transatlantic communications projects.

6 THE NORTH ATLANTIC: TECHNOLOGY, THE ENVIRONMENT, AND LIBERALIZATION

In 1988, BT ran an advertisement to promote its international communication services. The same year, BT and AT&T, in collaboration with France Télécom, laid TAT-8, the first digital, fiber-optic, transatlantic communications cable. The advertisement opened with a shot of an empty business office, a desk and chair on the right and a computer terminal on the left. A businessman picked up a Filofax and began to dial his telephone. The camera then pulled out to reveal that the shot came from outside the businessman's office window. Evoking the short film *Powers of Ten* by Charles and Ray Eames, the camera pulled out farther until all London, including BT Tower, were in view, and continued pulling out: the UK, Europe, and, finally, Earth, floating in space. A communications satellite flew in front of the camera, momentarily obscuring the view of Earth, until the camera, quickly, began to zoom back in, this time to an office in Manhattan, New York. Another businessman picked up his phone and answered the call. The closing title card displayed "British Telecom International" as a voiceover said, "It's you we answer to," referencing BT's privatization in 1984, advertised as "a public company goes public."[1]

One year later, in 1989, AT&T, the US telephone operator, also ran an ad promoting their international communication services. In contrast to BT's, this ad mentioned TAT-8, with an early sequence in the commercial showing the cable-laying. Using similar imagery to BT, however, the AT&T ad opened and closed with a shot of the Earth from space. After the cable-laying sequence, the ad shifted to AT&T's new information services, through which customers could access the "worldwide intelligent network." This sequence was interspersed with images of computers and space-age motifs, from satellites and earth stations to a NASA-style communications and telemetry control center, culminating with a phone call to astronauts in the space shuttle. The ad's closing title card contained AT&T's then-slogan, "The right choice," nodding to AT&T's divestiture in 1982, which broke the company into regional telephone providers and opened the long-distance US telephone network to competition.[2] These ads show that, in the wake of their monopolies ending, both BT and AT&T advertised their transatlantic communications services by focusing on satellites, rather than the cables that carried these services. This chapter investigates how the Post Office, BT, and AT&T managed this relationship between satellites and cables as transatlantic communications technologies. In doing so, the chapter shows how the Post Office and BT's turn from nationalized industry to private corporation intersected with the digitalization and liberalization of international communications markets.

The history of telecommunications is as much a history of international telecom infrastructures, regulations, and organizations as it is national. From the mid-nineteenth century, many telegraphic systems-builders prioritized international submarine telegraphy, particularly as a tool of empire.[3] The 1866 transatlantic telegraph led to Anglo-American hegemony over global telegraphy that has shaped contemporary discourse about global communications to the present day.[4] With the invention of wireless telegraphy, attention in the early twentieth century turned to the possibility of a global wireless network that, with some success, bypassed Anglo-American cable hegemony.[5] During the Cold War, communication satellites became the next focus of international communications. Led by the US, satellites became an extension of Cold War diplomacy that made international telephony and television broadcasting commonplace.[6] International communications has thus been a key site for making and remaking visions about the relationships among nations, corporations, empire, and the global economy.[7] But, in large part, this history has been dominated by the international agreements and organizations that govern and regulate these networks. The International Telecommunications Union, the key international governing body for telecommunications, originated as a "capitalist compromise" between nation-states and business.[8] It has since become a key site of "technocratic internationalism," where engineers weigh in on national and international issues and reach decisions about international telecommunications.[9] In the postwar era, as satellites took off, a new organization, INTELSAT, appeared to administer and govern international satellite communications.[10]

But focusing only on these international organizations would miss the central role that international communications markets played for national telecom operators such as the Post Office, BT, and AT&T. For much of the twentieth century, international telecommunications markets were quite well-regulated, particularly the transatlantic markets that this chapter focusses on. In Europe, the domestic PTT monopolies usually controlled their end of international communications links, which were either long-distance radio or submarine cables. In the US, AT&T was forced to divest its international subsidiary, ITT, in 1925 and could operate only the US end of international telephony links, while various carriers, such as RCA and Western Union International, competed to provide international telegraphy services. In general, telecom operators from one country, such as the Post Office or AT&T, could not operate telecom networks in another country. In the period that this chapter covers, various changes challenged the regulation of international communications markets.[11] Advances in satellite and submarine cable technology expanded the supply of international communications circuits, causing the cost of international communications to drop. In turn, business users, the biggest purchasers of these international links—which traditionally cross-subsidized domestic residential users by paying higher prices for international communications—pushed for deregulation and liberalization so they could pay lower prices and expand their international networks. Meanwhile, the deregulation of national telecom monopolies, especially of AT&T and BT, freed those companies to compete in offering new international network services. This all means that the North Atlantic telecom markets played a central role in the privatization and liberalization of both international and domestic telecommunications operators. In short, to understand the relationship between digitalization and the market turn in British telecommunications, understanding this transatlantic history is essential.

This chapter treats the North Atlantic as a "technological zone."[12] A technological zone is a space made through the ways that technologies connect and encircle firms and nation-states. For example, Windows compatibility creates a technological zone for software that spans nations yet remains bounded by one business, Microsoft. In keeping with this chapter's aim to understand how the denationalization of Britain's telecom infrastructure projected itself beyond Britain, these zones problematize the idea that there is a clear "inside" and "outside" to the nation-state. Because they involve malleable technologies, these zones require "frequent maintenance work."[13] Part of this work is institutional, undertaken by international organizations and transnational business partnerships. But it is also environmental. This history builds on work that draws attention to the historical intersections of technology and the environment, particularly the environmental history of the ocean, outer space, and communications technologies.[14] This chapter pays attention to the environments of North Atlantic communications, alongside its technologies and its institutions, because all three are essential to understanding the political economy of transatlantic communications that AT&T, the Post Office, and BT built and rebuilt from the 1950s to the 1980s.

This chapter unfolds in three parts. The first, which includes "Conquering the Atlantic" and "Hostile Environments," looks at how the national monopolies of the Post Office and AT&T worked on two new transatlantic communications projects, the TAT-1 submarine telephone cable and the Telstar communications satellite. This part explores both the environmental and regulatory histories of these projects, showing how these histories set up particular ways of thinking about the roles that cables and satellites played in transatlantic communications. Next, "The Single World System" looks at the "battle of the systems" between the North Atlantic cable system and a new international satellite system, INTELSAT. INTELSAT institutionalized Cold War satellite techno-diplomacy as part of US foreign policy agenda, while the North Atlantic telecom monopolies developed new organizational and environmental strategies to protect cables from the threat posed by satellites. Finally, "Cables Orbit Satellites" looks at how BT and AT&T, as newly denationalized telecom corporations, promoted new satellite and cable technologies, including TAT-8, the first digital fiber-optic transatlantic communications cable, to support the liberalization of international communications.

CONQUERING THE ATLANTIC

On September 25, 1956, AT&T and the Post Office opened TAT-1, the first transatlantic telephone cable. TAT-1 incorporated various new and old techniques for surviving the Atlantic environment. Crossing the far north of the Atlantic from Oban in Scotland to Clarenville in Newfoundland, the cable route was chosen for infrastructural and environmental purposes. Further south were telegraph cables, which might disrupt the new cable, and dangerous areas of the seabed, susceptible to turbidity currents, sediment-laden flows of water that could snap cables, as had happened following the 1929 Grand Banks earthquake and the 1954 Orleansville, Algeria, earthquake. On a technical level, new undersea repeaters, developed at Bell Labs, had made TAT-1 possible. These repeaters, which amplified and extended telephone signals, had to work reliably under the immense pressure at the bottom of the Atlantic. The repeaters used this high-pressure environment to their advantage. Their metal casings, unavoidably deformed during the cable-laying process, relied on the immense pressure at the ocean floor to pressure them back into their correct shape.[15] The Post Office had contributed to TAT-1 by designing its shallow-water repeaters, used in the link between Newfoundland and the North American mainland, and by using Her Majesty's Telegraph Ship, *Monarch*, the largest cable ship afloat at the time. TAT-1 used two cables, one for each direction of transmission, and *Monarch* was the only ship capable of transporting the entire cable length for one direction. TAT-1's final novelty, showing another intersection of technology and environment, was its innovative polyethene cladding, used to resist biological attack from marine bacteria, in contrast to previous cables' weaker polyvinyl chloride coatings.[16]

These harsh environmental conditions formed a major part of AT&T's publicity about transatlantic telephony in the late 1950s and early 1960s. Ads in boys' magazines talked about how the sea "could make a 'meal' of telephone cables" (figure 6.1) and explained Bell Labs' "experimental ocean," used to test cable specimens in saline conditions.[17] Advertisements also targeted business audiences, with ads in *American Banker*, the *Wall Street Journal*, and *Fortune* explaining how AT&T's "stormproof" Atlantic cable would allow them to expand US business interests in Europe.[18] A series of ads by AT&T called "Tele-Facts" deployed militaristic language to describe TAT-1 and, by extension, AT&T as "conquering the Atlantic." This militaristic tone pervaded many of AT&T's TAT-1 ads.[19] The same ads that ran in *American Banker* and *Fortune* explained how TAT-1 would be of "far-reaching value in national defense," while articles in *Bell Telephone Magazine* compared TAT-1 to Cold War projects like the Distant Early Warning Line and the Ballistic Missile Early Warning System.[20]

These Cold War geopolitical concerns were particularly evident in the trilateral negotiations for TAT-1 among the United States, Britain, and Canada.

THE SEA COULD MAKE A "MEAL"
OF TELEPHONE CABLES!

The sea has a billion "teeth" — the countless marine borers and bacteria which feed on organic materials in the deep. They also attack the great telephone cables laid to England, Hawaii, Alaska and Cuba, and are capable of doing enormous damage. In fact it has been discovered that some borers are capable of gnawing through thick lead!

Developing undersea telephone cables that borers and bacteria couldn't harm was a major undertaking of the Bell System. Before a foot of cable was laid, many tests were conducted to find insulation that could successfully resist the myriad teeth of the ocean.

Now, with more cables being planned, tests are continuing to find even lighter, stronger, more resistant substances with which to sheathe the cables. Some of these tests are in the ocean itself, some under controlled conditions at Bell Telephone Laboratories.

Battling the borers and bacteria of the deep sea is part of our job of providing you and your family with dependable, low-cost telephone service — whether you're calling across town or across the ocean.

 **BELL TELEPHONE SYSTEM**

19–2–854-1958–16 ins.–4 5-8 x 8–Boys' Life, Sept.; Junior High School Scholastic, etc., Oct. 3–
N. W. Ayer & Son, Inc.

FIGURE 6.1

The sea could make a "meal" of telephone cables! Credit: N W Ayer Advertising Agency Records, Archives Center, National Museum of American History, Smithsonian Institution.

TAT-1's route, devised by AT&T and Post Office officials, also fulfilled a British goal of strengthening UK–Canadian communications and extending the "all-red" Commonwealth communications route to reach New Zealand via Canada and the Pacific.[21] US officials found this problematic and had two security concerns. First, regarding the cable landing in Canada rather than in the United States, and second, over the plans for the Post Office rather than AT&T to design and contract out construction of the shallow-water Newfoundland–Nova Scotia section.[22] The US proposed instead to staff the Canadian cable stations with US AT&T staff. In response, Canadian officials expressed concerns on security and commercial grounds, fearing that it would pave the way for the commercial expansion of US telecommunications into Canada.[23] The Canadians' attitude raised concerns for the UK, where Foreign Office and Post Office officials worried that Canadian intransigence would cause the US to lay a cable directly to France instead.[24] The resulting compromise was that a Canadian AT&T subsidiary, the Eastern Telephone and Telegraph Company, would operate Canadian sections, the Canadian Overseas Telecommunication Corporation would take a 10 percent minority stake in TAT-1, and the Post Office would design the Newfoundland–Nova Scotia section. In return, the next transatlantic cable, TAT-2, would run from the United States to France to avoid concentrating traffic through the UK.[25]

An AT&T publicity film for TAT-2, which was laid in 1959, further shows the entanglement of these cables' environments with US Cold War geopolitical concerns about Europe. The film opens with scenes of waves crashing on rocks and emphasizes the cable's victories over the "many-mooded sea," describing battles against the wind, cold, and icebergs. It concludes that the cable "should do much to bring many nations closer together, both politically and economically, and contribute significantly to the defense needs of the free world" and was "man's newest memorable victory over distance and the sea."[26] This film captures how the undersea cables' environments were crucial to a discourse about extending US military and economic influence into Europe. The early transatlantic telephone cables were part of the US "consensual hegemony" over Europe, in which the US used scientific and technological projects in the early Cold War to aid European reconstruction and serve its Cold War defense interests.[27]

In contrast, the Post Office used TAT-1's environment to emphasize the British scientific and technological ingenuity that had made TAT-1 a "world first." At the cable's opening ceremony, Charles Hill, the postmaster-general, highlighted the engineering prowess and patient research behind the cable, while the Post Office's official souvenir booklet emphasized British oceanographic knowledge and manufacturing skill.[28] Gordon Radley, then the Post Office's director general, spoke on the BBC radio Home Service program *Science Survey* in September 1956 about TAT-1 as a "significant scientific achievement."[29] Radley described the cable resting in the "perpetual darkness and ooze of the sea bed," evoking Rudyard Kipling's poem "The Deep-Sea Cables": "There is no sound, no echo of sound, in the deserts of the deep, / Or the great grey level plains of ooze where the shell-burred cables creep."[30] But where Kipling's poem portrayed cables as a globally unifying force, transcending their environment, Radley's talk and the Post Office's TAT-1 publicity instead resembled the nationalism of Highgate Wood. Even though AT&T's deep-sea repeaters had enabled TAT-1, the Post Office's repeated emphasis on Britain's contributions revealed a distinctly nationalist tinge.

This British nationalism also extended to the Post Office's efforts to unsettle the regulation of transatlantic communications in the UK's favor.[31] Before TAT-1, transatlantic telephony and telegraphy were operated under different regulatory arrangements. AT&T and the Post Office operated radiotelephony links, while US "international record carriers," such as Western Union International and RCA, operated wireless and cable telegraphy. This was a regulatory arrangement set up in the US in the early twentieth century to prevent AT&T expanding its domestic monopoly to international communications. But TAT-1, as the first coaxial transatlantic cable, would have a bandwidth that meant that the Post Office could offer both transatlantic telephone and telegraph services. This threatened the balance of power in the US between AT&T, as the international telephony provider, and the record carriers, which provided international telegraphy. Soon after the Post Office and AT&T signed the first agreement to lay TAT-1 in 1952, ITT thus announced that its subsidiary, the Commercial Cable Company, would lay a new coaxial transatlantic cable, code-named Project Deep Freeze, for telegraph services alone. This cable upheld the existing US regulatory framework and would continue to exclude the Post Office from transatlantic telegraphy. After negotiations that went on into 1956, the Post Office continued to refuse ITT a license to land Deep Freeze in the UK, while the US continued to block telegraph services over TAT-1.

When TAT-1 launched, it thus carried telephony only between the US and the UK, but, because it went via Canada, the Post Office used TAT-1 for international telegraphy services to Canada. TAT-1 was a huge success. It carried twice as many calls as radiotelephony had done in the previous year and brought in significant revenue for its US, Canadian, and British operators. Indeed, TAT-1 was so successful that the UK and Canada immediately began planning a new transatlantic coaxial cable, CANTAT-1, which would carry transatlantic telephony and telegraphy and would exclude AT&T. This finally forced changes in the US regulation of international communication. The US international record carriers could not abide the chance that their North American customers would start routing all telegraph traffic via Canada to go over CANTAT-1. The FCC thus agreed, in 1959, that the record carriers could lease circuits from AT&T over TAT-1 and the Canadian Overseas Telecommunication Corporation over CANTAT-1, finally ending the separation between international telegraphy and telephony. This meant that these international record carriers stopped laying new transatlantic cables, and indeed, through the 1960s, they also stopped operating their old transatlantic telegraph cables, instead leasing circuits on coaxial telephone cables, which, if they landed in the UK, were all partly owned by the Post Office. The Post Office thus achieved its goal of excluding the US record carriers from the UK, and it had expanded its monopoly over the British end of transatlantic communications from telephony to telegraphy.

HOSTILE ENVIRONMENTS

In the early 1960s, space-based communications took off. AT&T and the Post Office initially presented satellites as complementary parts of their international services, although new attitudes to space as an environment and zone of Cold War conflict also appeared. In the US, this occurred with the launch of AT&T's Telstar satellite, while in Britain, this happened with the Post Office's construction of Goonhilly Downs, Britain's first satellite earth station. Telstar, launched in July 1962, was the first satellite to relay telephony and television across the Atlantic, from AT&T's earth station in Andover, Maine, to British and French earth stations at Goonhilly Downs, Cornwall, and Pleumeur-Bodou, Brittany. Bell Labs' initial research into satellite telephony came in 1955, but its R&D program began in earnest in 1959 when AT&T agreed with NASA that AT&T would design and construct an active communications satellite for NASA to launch.[32] This satellite, Telstar, was roughly spherical, composed of seventy-two facets covered in sixty solar cells and three mirrors, which aided satellite tracking from Earth. The satellite weighed 170 pounds and contained a single amplifier that could transmit a wide-band signal, such as a television broadcast, one way, or two narrow-band signals, such as a telephone call, two ways. Telstar's purpose was not only to prove the viability of satellite communications but also to gain an understanding of the space environment, particularly the Van Allen radiation belts surrounding Earth, discovered by James Van Allen at the University of Iowa in 1958 using data from the Explorer 1 and Explorer 3 satellites.[33]

AT&T situated these space activities alongside transatlantic telephone cables in its publicity. After its early experiments in space communications in the late 1950s, AT&T ran a widely published series of ads with the header "From Beyond the Sky to Beneath the Seas" in military and science magazines (figure 6.2), juxtaposing the sea and space environments to demonstrate the breadth of AT&T's accomplishments.[34] The ads' appearance in military magazines further demonstrates the militarization of Cold War transatlantic communications discourses. College recruitment ads also used TAT-1 and Telstar. One poster described how "Between Outer Space and the Deep Sea There's a Wide Range of Opportunity in the Bell Telephone Companies," while another explained how "progress in the Bell System," among other things, "swims" and "orbits."[35]

The Telstar experiment also interlinked the hazards of the space environment with the growing environmental awareness of the 1960s, which raised concerns about US militarization of the space environment after the Telstar experiment. Telstar had been launched not only as a communications satellite, and, as AT&T publicity explained, it was also a "space laboratory," "operating in the unknown environment of hostile radiation and micrometeorite dust," sending back data about the space environment to Bell Labs.[36] The day before Telstar launched, the US detonated Starfish Prime, the largest man-made nuclear explosion in outer space, part of a series of high-altitude nuclear weapons tests called Operation Fishbowl. This detonation energized the Van Allen belt, which damaged transistors on Telstar, causing it to fail. The failure of Telstar and seven other satellites, including Ariel I, Britain's first satellite, caused by Starfish Prime, highlighted the hazardous environment of space and fed environmental concerns about the damage US military programs were doing to outer space. Newspaper articles linked Telstar's failure to Operation Fishbowl's potential damage to the space environment, and James Van Allen criticized the military tests, which used data

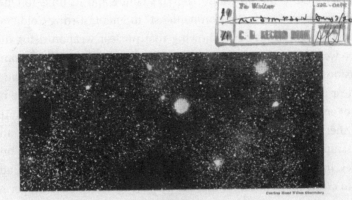

Courtesy Mount Wilson Observatory

FROM BEYOND THE SKY
TO BENEATH THE SEAS

In the field of communications, two extraordinary events have occurred within a short span of time. One was the linking of Europe to America by the submarine telephone cable. The other was the sending of radio signals from U. S. satellites in outer space.

Both achievements depended on developments from Bell Telephone System research. The cable was made possible by the development of long-life electron tube amplifiers which are able to withstand crushing pressure on the ocean floor. The satellites derive their radio voices from transistors—products of basic research in semiconductor physics.

All of this basic research is done for the Bell System by Bell Telephone Laboratories. Here, over 3000 professional scientists and engineers explore and develop in physics, mathematics, electronics, chemistry, mechanical engineering, even biology — in every art and science which can help improve electrical communications.

Through this work at the Laboratories, the Bell System has helped make your telephone service the world's finest — and will keep it so.

BELL TELEPHONE SYSTEM

M—3-666-1956-1 page—7 x 19—Army, Navy, Air Force Journal, etc., Oct.—N. W. Ayer & Son, Inc.

FIGURE 6.2
Telstar and TAT-1: "From beyond the sky to beneath the seas." Credit: N W Ayer Advertising Agency Records, Archives Center, National Museum of American History, Smithsonian Institution.

from Telstar to study the explosions before its failure, for projecting a "sinister" air around the program.[37] Telstar's failure and its links to Operation Fishbowl were a "proto-environmentalist" moment, stirring Cold War environmental insecurities and showing that nuclear weapon detonations in space were not just about weapons testing, but also about environmental transformation.[38] Telstar's failure, which made visible the militarization and nuclear pollution of the space environment, was thus part of the broader rise of environmentalism as "a child of the Cold War."[39]

Other early approaches to space communications showed how attempts to manipulate the space environment came from efforts to bypass undersea cables. In 1961, MIT's Lincoln Laboratory began Project West Ford, attempting to create an artificial ionosphere by placing 480 million copper needles in orbit. These needles would act as a passive antenna to bounce communication signals from one place to another. Project West Ford aimed to reduce the US military's reliance on undersea cables after a Soviet fishing trawler was suspected of deliberately cutting transatlantic telephone and telegraph cables owned by AT&T and Western Union in 1959.[40] Protests against Project West Ford occurred in both the UK and the USSR, and the project eventually came under criticism within the US too. British radio-astronomers, such as Bernard Lovell, worried about how the needles might affect radio astronomy, while in the USSR, *Pravda* attacked the US with the heading "USA Dirties Space," calling the needles "space junk."[41] In the US, the *New York Times* argued that the US had no unilateral right to influence the space environment.[42] Operation Fishbowl and Project West Ford show that, while AT&T initially positioned Telstar as complementing TAT-1, early space communications escaped this rhetorical frame, instead representing the US militarization of the space environment.

Meanwhile, the Post Office used space communications to emphasize British engineering ingenuity and technological mastery over the natural environment. Satellite histories tend to focus on the cosmic and not the terrestrial, but satellite communication is more than satellites in space.[43] It is also a vast, material, terrestrial infrastructure composed of, by now, hundreds of earth stations around the world. Earth stations are essential nodes in communication satellite infrastructure, and the Telstar experiment required three earth stations in three different countries. The Post Office built its first earth station, Goonhilly Earth Station Office, in 1962 on Goonhilly Downs, an isolated, elevated plateau on the Cornish peninsula with broad sightlines.

The Post Office pursued a unique direction for Goonhilly's design. The first antenna, Antenna One, also known as Arthur, was the world's first satellite communication antenna with a parabolic design. The antenna was designed by Charles Husband, the engineer behind Jodrell Bank's Lovell Telescope, the world's largest steerable radio telescope, which had also used a parabolic design and was the world's first satellite "dish."[44] The Post Office proudly touted Antenna One's parabolic design as a uniquely British design concept that did not need environmental protection, in contrast to AT&T's Andover, Maine, earth station, which utilized a "horn" antenna that required protection from the environment by a distinctive "golf ball" protective radome.[45] The Post Office mobilized Goonhilly's dish as part of its publicity, featuring it in its "Progress" poster series (figure 6.3).[46] The Post Office later proudly touted how the British parabolic design became the template for subsequent earth stations worldwide, again emphasizing the Post Office's role in British technological exports. Goonhilly, however, was not a complete success. Initial communication with Telstar failed because a component was accidentally inverted, disrupting the Post Office's image of British technological sophistication. The prime minister, Harold Macmillan, demanded an explanation from the Post Office, which explained the simple error behind Goonhilly's failure while also replying that Goonhilly had cost a quarter as much as the French earth station and had showcased Britain's expertise in antenna construction.[47]

The expansion of space communications brought concerns about interference in the radio spectrum, which had consequences for Goonhilly. As international satellite communication developed alongside domestic microwave networks, the radio spectrum became increasingly congested. Radio astronomers at Jodrell Bank had already experienced such issues, which were a familiar problem for scientific establishments, where electrical interference often disturbs instruments.[48] Mitigating interference had thus been an early priority for the Telstar experiment. In order to standardize communications and replicate signal transmission, the French had, at considerable expense, duplicated and imported AT&T's Andover earth station for their earth station at Pleumeur-Bodou, Brittany.[49] For Goonhilly, growing interference meant that, by the 1970s, it could no longer serve as Britain's only earth station. The Cornish peninsula had been ideal for transatlantic satellites, but southeast facing aerials, pointed at satellites stationed above the Indian ocean, were prone to interference from French microwave networks across the English Channel (figure 6.4).[50]

FIGURE 6.3

Goonhilly also appeared in the "Progress" poster series. Source: TCB 420/IRP (PR) 1, BT Archives. Courtesy of BT Group Archives.

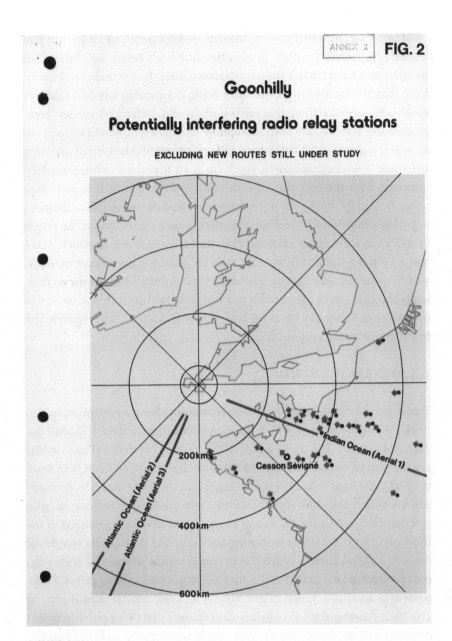

FIGURE 6.4

Interference between Goonhilly and French microwave relay stations. Source: TCC 55/6/145, BT Archives. Courtesy BT Group Archives.

Britain's second earth station, Madley, which opened in 1978, thus had a northern, inland position in Herefordshire with better sightlines. This location protected Madley from interference from the continent and Goonhilly, which remained in service until 2008. Goonhilly's fate and Madley's construction demonstrate that earth stations and, indeed, all wireless communication stations, are both susceptible to interference and producers of it, which is an essential detail for a full environmental history of the radio spectrum. The electromagnetic spectrum is an invisible environment that intersects with the spatiality of communications infrastructure and other scientific and technological institutions such as radio telescopes and metrological laboratories. By 1968, this was already such a concern that an article in AT&T's *Bell Telephone Magazine* described the spectrum, in which AT&T had invested a significant amount with its domestic microwave network, as a "natural resource" being "polluted."[51] Goonhilly's history shows that, while the Post Office presented it as a uniquely British earth station, it was by no means nationally bounded and instead remained always open to the transnational extension of the electromagnetic environment.

THE SINGLE WORLD SYSTEM

Telstar proved the viability of satellite communications, so attention turned to global satellite systems, which ultimately took form in the INTELSAT "single world system." The US spearheaded the creation of INTELSAT as a satellite system with global access and Cold War objectives. Soviet success with Sputnik and Yuri Gagarin had left the US lagging in the space race. The Kennedy administration thus saw an international satellite alliance as a way to gain prestige, catch up to the USSR, and align neutral developing nations in the third world with the US. Broadcasting television and propaganda worldwide through a global satellite system could be a powerful weapon for US foreign policy. Negotiations thus started with foreign governments to gather a consortium of nations to invest in and support an international satellite system. The 1962 Communications Satellite Act created COMSAT, a publicly traded corporation that represented the US in these negotiations, and the courtship of foreign governments began.[52]

INTELSAT had a profound influence on the Post Office, BT, and AT&T's development of transatlantic communications. The Post Office had to negotiate its commitment to submarine cables with the growing support for a single international satellite system. The Post Office's early interest in satellites came from its desire to create a complementary Commonwealth satellite system that would interlink with cables.[53] There was also potential for a European satellite system, in which the Post Office was reluctant to invest. It would have to pay both purchase costs for the European launcher, developed by the European Launcher Development Organisation, and development costs, as the launcher was still in development. The Foreign Office foresaw political problems with both Commonwealth and European systems and, given the US head-start on satellite development, convinced the Post Office that the US-led international system represented the best option for British industrial, telecommunications, and Commonwealth interests.

The Post Office enrolled the Commonwealth in this strategy through the Commonwealth Conference on Satellite Communications in 1962. While the Post Office still emphasized at the conference its preference for a complementary cable and satellite system, it led talks that concluded that the Post Office, in conjunction with Canada, should undertake exploratory talks with the US about the new international satellite system.[54] European interests coalesced through the Conference of European Postal and Telecommunication Administrations, which agreed not to establish a regional satellite system in opposition to the US, although the European Conference on Satellite Telecommunications, CETS, was established for long-term planning. Through concerted negotiations by COMSAT and the US State Department, the groundwork for a single global satellite system had been laid by 1963.

A significant opportunity to voice further opinions came in 1963, at the Extraordinary Administrative Radio Conference, organized by the International Telecommunications Union. The US used this conference to secure frequency allocations for the single global system, strengthen ties with its European allies, and undertake "missionary" work, promoting technical assistance programs with potential third-world supporters.[55] Meanwhile, the Post Office used it to parade Goonhilly and argue in favor of an interconnected satellite-cable system.[56] In a technical paper, the Post Office again touted Goonhilly's unique, parabolic, unprotected antenna and described another unique feature, its use of computer prediction, rather than automatic tracking, for the steering aerial. This computer used orbital data to predict satellite movement so that the antenna would not need to be steered to acquire satellites after they appeared over the horizon. The Goonhilly computer proved helpful at the conference and was used, via Telex, to calculate degrees of interference between different radio-communication services, which was another opportunity for the Post Office to deploy Goonhilly to emphasize its technological prowess.[57] The Post Office also submitted a resolution that a satellite system should interconnect with international communication cables. The resolution argued that the time-delay and Doppler frequency shifts associated with satellite transmission necessitated an interconnected system that provided various transmission routes, such as undersea cables, to mitigate these issues. This resolution and the continuing commitment by countries such as France to their cable networks meant that the US eventually conceded, putting sharing criteria in place for satellite systems to interconnect with terrestrial networks.

In 1964, INTELSAT finally formed through an interim agreement that would formalize in 1971. The US, through COMSAT, owned half, while major partners from Europe and around the world, including Canada, Japan, and Australia, owned the other half. As Britain's representative, the Post Office was the second-largest single shareholder behind the US, with an 8.4 percent ownership share.[58] Reflecting its goals and organization as a "single world system," INTELSAT supported a globalizing liberal democratic Cold War discourse through the 1960s and 1970s and emphasized satellites as superior and environmentally transcendent compared to submarine cables as part of this discourse. "One world" discourses of various valences proliferated through the 1960s. In 1962, the media theorist Marshall McLuhan, in *The Gutenberg Galaxy*, popularized the term "global village," while the digital utopian writer-entrepreneur Stewart Brand seized on the first photos of Earth from space as a call-to-arms for his countercultural publication, the *Whole Earth Catalog*.[59] Before a specific INTELSAT discourse stabilized, the first INTELSAT satellite, Early Bird, was caught up in a McLuhan-esque "communications explosion" discourse. For example, Early Bird appeared on the front cover of *TIME* magazine, drawn by Saul Steinberg. Steinberg drew Early Bird beaming a "communications explosion," an unsettling jumble of geometric shapes, into a man's head, designed to convey the view of space communications as "a maze of reflections of one thing to another" and "the somewhat frightening prospect of man's new capability to store a mass of information and, on signal, send it to anywhere in the world."[60]

By the end of the 1960s, the INTELSAT discourse had stabilized into a rhetoric of communication satellites as agents of global peace and unity, with registers of liberal democratic capitalism and the highly anticipated information revolution. An ad by Hughes, Early Bird's manufacturer, explained that with the satellite, "the future looks bright. It includes increased world trade and better understanding between nations."[61] A COMSAT fact sheet explained how satellites would "increasingly handle even more futuristic chores" such as data exchange and facsimile. At the Early Bird inaugural address, President Lyndon B. Johnson proclaimed that the Early Bird service "brings closer together lands and people who share not only a common heritage but a common destiny."[62] Two years later, in 1967, at the launch of the INTELSAT II satellite over the Pacific, Johnson invoked similar rhetoric. He described that satellites would make space "a zone of peace, devoted to the purposes of all mankind," while Rosel Hyde, chairman of the Federal Communications Commission, described how the satellite would improve "the flow of knowledge and commerce across new high-capacity highways of communications."[63]

This discourse peaked at the signing of the official INTELSAT accords in Washington in 1971 with the contributions of science-fiction author Arthur C. Clarke. Clarke had originally proposed a geosynchronous satellite system in a 1945 article in *Wireless World*, and as such, INTELSAT regularly invoked his fame and predilection for grand predictions. At the accords' signing, Clarke explained his belief that "communication satellites can unite mankind" and informed the signatories that they had "just signed a first draft of the Articles of Federation of the United States of Earth."[64] In a 1971 article for *Popular Science* about the new INTELSAT IV series of satellites, Wernher von Braun, the infamous German American aerospace engineer, wrote that INTELSAT would establish what "Arthur Clarke, prophetic writer on space, has called 'mankind's first nervous system,' which will 'link together the whole human race.'"[65] The INTELSAT III series press handbook used the same quote. Clarke also wrote a guest editorial in *Bell Telephone Magazine*, in which he linked communication satellites with computers, suggesting that, while the enormous channel capacity of satellites may not be needed for a billion simultaneous human conversations, they would certainly be needed for computers, "which are becoming more talkative than their human creators."[66] What this discourse missed, however, was the extent to which US culture shaped INTELSAT's "one world." INTELSAT satellites monopolized international television transmissions, which were dominated by the broadcast of US television abroad. These satellites were thus less about "one world" and more about the US following its postwar "consensual hegemony" over European reconstruction with a cultural hegemony over international media.[67]

INTELSAT reorganized international communications to favor satellites over cables, emphasizing satellites' large bandwidth, as well as undersea cables' fragility and satellites' apparent environmental transcendence. Wernher von Braun's *Popular Science* article touted the superior capacity of INTELSAT IV compared to the "puny" capacity of the "most sophisticated transatlantic cable." A COMSAT brochure titled "New Communications Era" explained that the "archaic" cable system was no longer necessary and that Early Bird nearly doubled the capacity of TAT cables at less than one-fifth the cost.[68] In June 1965, COMSAT seized on the failure of the Canada-to-England transatlantic cable, CANTAT, to petition the FCC for temporary replacement service via Early Bird, and quickly publicized Early Bird's rescue of transatlantic communications.[69] The same occurred three years later when COMSAT publicized how INTELSAT satellites had carried their heaviest ever load of Atlantic traffic after two transatlantic cables had been damaged.[70] An INTELSAT educational booklet explained how satellites were superior to cables both as agents of global peace and understanding and as a medium for many new types of communication.[71] INTELSAT also perpetuated the notion that satellites could escape the "inherent limitations" of the environment, whereas terrestrial communications, in the environmental degradation of radio communications or the fragile materiality of cables, could not.[72] These ideas were also articulated outside INTELSAT and COMSAT. *Aviation Week* reported INTELSAT IV's capacity of three thousand to nine thousand circuits compared to the 750 of the most recent transatlantic cable, while *TV Guide* drew together the supposed differences in capacity and environment in an article about the growing demand for international communications.[73] *TV Guide* simultaneously emphasized the superior capacity of INTELSAT and the environment of submarine cables with the rhetorical question, "Meet that demand with undersea cables? They'd drown in an ocean of words. But satellites can handle it." Cables had been submerged, both literally and figuratively, while satellites appeared a capable, transcendent technology.

CABLES ORBIT SATELLITES

The INTELSAT system and communication satellites significantly influenced the Post Office, BT, and AT&T's development of transatlantic communications. This influence unfolded in three ways. First, the INTELSAT system inspired a North Atlantic Systems Conference, led by AT&T and the

Post Office, that resisted the satellite system. Second, BT and AT&T began to develop systems that would bypass INTELSAT and pave the way for the liberalization of international communications. Third, BT and AT&T subverted INTELSAT's discourse about satellite communications to advertise their own corporate, privatized model of international communications, focusing on satellites, computers, and free enterprise.

The Post Office continuously researched and monitored the proficiency of both satellites and submarine cables. In 1968, researchers produced five reports comparing satellites and cables.[74] One paper addressed noise performance, concluding that cable circuits had marginally better performance, while satellites were more susceptible to rain and atmospheric conditions causing bursts of noise. One compared propagation conditions while another addressed the fallibility of earth stations, noting that snowfall on a German earth station's radome had canceled a satellite TV broadcast from Germany, and that radome repairs had also put Andover and Pleumeur-Bodou out of action for extended periods. The final paper compared satellites and cables' relative secrecy, concluding that submarine cables were more secure, but also noted that in the future, satellites with highly directional aerials could target just a few square miles around the earth stations, increasing security. Donald Wray, the deputy director of engineering and the Post Office engineer who had planned Goonhilly, dryly noted, however, that if "the Red Chinese started building an earth station in Cornwall their activity would not pass unnoticed." Wray's overall analysis emphatically concluded that cables were superior to satellites. Cables had greater secrecy, simplicity, lifespan, and transmission time, while satellite earth stations were more complex and had higher personnel requirements than cable stations. The Post Office's Joint Submarine Systems Development Unit, run with Cable & Wireless, also noted that satellites were less susceptible to malicious and electrical interference but pointed out that, in the event they were damaged, cables could be repaired whereas satellites could not.[75]

By 1976, eight cables crossed the Atlantic, six in the TAT series and two in the CANTAT series, and various techniques were developed to protect and repair these cables. In 1970, the Post Office used a "sea-plough," developed by AT&T, to bury 80 miles of TAT-3 off the Cornish coast to protect the cable from fishing trawler damage.[76] In 1970, I. R. Finlayson, the Post Office's submarine superintendent, commissioned a marine consultant, Lieutenant Commander Lovell-Smith, to report on the viability for a diving unit to repair submarine cables.[77] Finlayson also collaborated with the Marine Technology Support Unit at Wantage Research Laboratory, part of the United Kingdom Atomic Energy Authority, on developing underwater habitats in which engineers could repair submarine cables.[78] The Post Office never established a diving unit and apparently never deployed submersible habitats, but its interest in these strategies highlights the pressure to devise new ways of quickly repairing damaged cables. One successful strategy was the use of submersibles. In the early 1970s, the Post Office used manned submersibles, called Pisces, to bury and repair cable, and later used two remotely controlled submersibles, known as SCARABs, for submarine maintenance work.[79] Another Post Office project developed a new grapnel for cutting deep-sea cables and bringing them to the surface for repairs. By 1979, Martlesham Heath had developed the "cut and hold grapnel," which could simultaneously cut a cable and lift it for repairs. The grapnel could work at depths of 5,000 fathoms, had a sonar surveillance system, and used a built-in power source to provide hydraulic operation. This new grapnel reduced grappling time by one-third and total repair time by just over one-fifth and was heralded as a leap forward in cable repair, quickly finding customers abroad.[80]

The maintenance of submarine cables required close cooperation and collaboration among North Atlantic telecommunications companies, which paved the way for more organized resistance by AT&T and the European PTTs, which favored cables, against INTELSAT. In 1975, a consortium of North Atlantic telecom companies, including the Post Office and AT&T, signed the North Atlantic Cable Maintenance Agreement, to pool funds and share resources over cable maintenance, beginning with the purchase of the SCARAB remotely controlled submersibles. The technical and diplomatic considerations that influenced cable planning, as well as the benefits of pooling resources, demonstrated by arrangements such as the North Atlantic Cable Maintenance Agreement, meant that, in 1977, the Post Office hosted the first North Atlantic Systems Conference.[81] The conference, composed of telecom administrations from Western Europe and North America, met in Eastbourne, Sussex, and was ostensibly transmission-neutral, claiming to discuss all communication links across the North Atlantic. As such, the conference also invited representatives from COMSAT and INTELSAT.

The telecom administrations, however, had an ulterior motive of reestablishing cables' place in transatlantic communications. In 1976, AT&T and the Post Office, along with CTNE, the Spanish telephone administration, discussed developing the next round of transatlantic cables, TAT-7 and TAT-8.[82] AT&T had found it challenging to secure FCC approval for transatlantic telephone cables, given the US commitment to communication satellites and the security of the COMSAT-INTELSAT diplomatic and legal instrument. The telecom companies thus aimed to give transatlantic telephone cable planning greater weight in the FCC's eyes. The result was the North Atlantic Systems Conference, which earned legitimacy by claiming system neutrality and having COMSAT and INTELSAT as participants. Somewhat too late, COMSAT's representative, Jack Oslund, realized that the conference was, in his words, an attempt to "INTELSATIZE the cable planning process to achieve a comparability with the satellite process in the eyes of the US government," by which he meant the FCC.[83] By this point, however, plans for TAT-7 had been approved, the groundwork for TAT-8 laid, and agreements made by the North American and European telecom companies for further conferences in the series.

In 1978, TAT-7, the last analogue coaxial cable in the TAT series, was laid, and ten years later, in 1988, TAT-8, the first digital fiber-optic cable, was laid. TAT-8 was a joint venture led by the newly privatized BT, the newly divested AT&T, and France Télécom, created in 1988 in preparation for the French separation of posts and telecoms, which in turn paved the way for liberalization and privatization. The cable cost £225 million, of which BT contributed £34 million, the second-largest share. The cable's novelty lay not just in its new transmission medium, optical fiber, nor its new digital transmission mode, but also its use of an underwater branching unit on the continental shelf off the British coast. This meant that the cable provided links from the United States to both Britain and France, as well as a cross-channel fiber-optic link.[84] TAT-8 could carry forty thousand simultaneous telephone calls, which was almost a tenfold leap over TAT-7's 4,200-circuit capacity and more than a threefold increase from the most recent INTELSAT series, INTELSAT V, which could carry twelve thousand calls.[85]

TAT-7 and TAT-8 arrived during a crucial period in the Post Office and AT&T's attempts to break INTELSAT's dominance over transatlantic communications. This became an especially important goal for AT&T and the Post Office as their own regulatory frameworks transformed. In 1982, the US federal government broke up AT&T's monopoly, requiring that AT&T divest its local networks into independent regional subsidiaries. In 1981, meanwhile, the Post Office became British Telecom, had its domestic monopoly turned into a duopoly, and then was privatized in 1984. Both, however, maintained their international services and networks, and deregulation encouraged them to expand further into the lucrative international telecommunications market. This came at a time when both the US and British governments joined AT&T and BT in wanting to see an end to INTELSAT's satellite monopoly over transatlantic communications.[86] The British government wanted to increase BT's appeal to international business users and believed that undermining INTELSAT might open up opportunities for British aerospace manufacturers, which historically had little success in securing contracts to supply INTELSAT. The US government, meanwhile, had seen its majority stake in INTELSAT declining as more and more countries had joined INTELSAT through the 1970s. Furthermore, INTELSAT's demand for communications satellites could not keep up with the supply capacity of the US aerospace industry, so liberalizing satellite communications would also benefit US satellite manufacturers.

In 1985, the FCC thus made several decisions that broke INTELSAT's dominance. In addition to authorizing TAT-8, the FCC approved further private transatlantic fiber-optic cables from new suppliers Tel-Optik and Submarine Lightwave. But, in perhaps the most surprising decision, the FCC also approved an application from Orion Satellite Corporation to place a private communications satellite over the Atlantic. In total, the FCC approved 330,000 circuits for transatlantic communications, more than tripling Intelsat's then-capacity of 100,000.[87] While the FCC's approval of the Orion satellite signaled a formal end to INTELSAT's transatlantic monopoly, the Post Office, AT&T, and other North Atlantic telecom operators had already successfully resisted this monopoly since the mid-1970s by INTELSAT-izing cable planning, so securing the FCC's approval of TAT-7.

Furthermore, while Orion was the first private satellite over the Atlantic, BT had already begun a service, SatStream, that showed how satellite liberalization could circumvent existing domestic monopolies. SatStream was an international data service that allowed customers to connect directly with each other via rooftop satellite dishes, which transmitted the signal from one rooftop, via an INTELSAT satellite, to another rooftop satellite dish. BT launched this service in 1984 to connect businesses in Canada, the UK, and Europe and called it an "integral part of British Telecom's network market strategy" as BT searched for new clients to remain competitive after liberalization.[88] Because SatStream was effectively a rooftop-to-rooftop service, it bypassed domestic telecom networks in Canada and Europe. This meant that BT's SatStream clients did business only with BT, which leased the satellite lines from INTELSAT, rather than having to route their traffic via their domestic telecom provider to a satellite earth station. SatStream was not a commercial success—by 1986, BT only had one SatStream customer—but it was an important proof-of-principle that the liberalization of satellite communications could be used to bypass domestic telecom monopolies.[89]

BT was, in effect, using satellites to evade the monopolies of its fellow European PTTs, possibly inspiring Orion's proposal to the FCC in 1985, which further privatized and liberalized transatlantic satellite communications.[90] BT's new network market strategy of prioritizing international business customers was also apparent in another development, London Teleport, an urban satellite earth station that opened for service on February 1, 1984, in London Docklands to bring "high-speed telecommunications to the fingertips of the City" using the City of London's new fiber-optic network.[91] Announcing that it was "bringing space-age communications to the heart of London," BT emphasized the teleport's business orientation by highlighting its videoconferencing capabilities, offering "the busy executive the ability to conduct real time, face-to-face meetings without the need to commit valuable time and resources to travel." These developments, especially SatStream, show how BT found new technologies, both cables and satellites, that could help it liberalize the regulation of transatlantic communications, just as the Post Office had done thirty years earlier with TAT-1.

BT made this a priority because of the massive expansion of international data communications markets, and both BT and AT&T showed this in their advertising to business customers. These ads underscored the importance of these international data communications markets by linking satellites, information technology, and international business. The BT International and AT&T ads mentioned above demonstrate how these corporations juxtaposed satellites, computers, and business, all in the same year that TAT-8 launched. In 1984, another BT International ad told customers that international communication was "uniting the business world," while a 1986 ad announced that international videoconferencing was businesses' "short cut to the global village," recasting McLuhan's concept of the global village into a capitalist vision of globalism.[92] Another 1986 ad publicized global data communication links as "The Information World."[93] AT&T's advertising in the 1980s also reflected this, presenting AT&T as "The Knowledge Business" and that, through its international services, "Bell Brings the World Closer."[94] A series of ads addressing AT&T's divestiture linked international services explicitly to the information age. In one, AT&T's CEO of overseas services, Morris Tanenbaum, explained that AT&T's global network was "the foundation for the information age."[95] This paralleled the AT&T TV ad described at the beginning of this chapter, in which AT&T promoted its "worldwide intelligent network." In a 1986 ad, "Issues of the Information Age: Promises Kept, Promises to Keep," which ran in the *Wall Street Journal*, AT&T explained how international communications was key to achieving a worldwide "Telecommunity," a "vast global network of networks, the merging of computers and communications."[96] This emphasis on satellites and data communications was not exactly new. Satellites had been described as a key information technology, heralding an "information revolution," by economists since the 1960s.[97] But what both BT and AT&T showed was that, in a world of expanding data communications, the economic importance of both satellites and cables was, for them, in how they could further the privatization and liberalization of international communications.

CONCLUSION

From the 1950s to the 1980s, the Post Office, BT, and AT&T developed new satellite and cable systems to increase their share of the revenue generated by transatlantic communications, and this shaped and was shaped by both international and domestic telecom liberalization. In the 1950s and early 1960s, the national monopolies of the Post Office and AT&T pioneered two new transatlantic communications systems, TAT-1 and Telstar. TAT-1 already showed how the Post Office used new coaxial technology, capable of simultaneously transmitting telegraphy and telephony, to take international telegraph markets away from US international record carriers. But transatlantic cables were soon overtaken by satellites. As satellites emerged, the Post Office worked at the 1963 Extraordinary Administrative Radioconference to advance a complementary communications agenda that favored connecting cables and satellites. Cold War interests motivated the US-led construction of a worldwide satellite system, taking shape as INTELSAT, while national telecom monopolies pursued various organizational and environmental strategies to safeguard cables' futures.

This led to a "battle of the systems" between INTELSAT's global system of geostationary satellites and the North Atlantic cable system. INTELSAT institutionalized Cold War satellite techno-diplomacy as part of the US foreign policy agenda. Meanwhile, North Atlantic telecom monopolies, led by the Post Office and AT&T, developed new forms of corporate partnerships to protect cables, leading to the "INTELSAT-izing" of North Atlantic cables to mirror satellite planning, successfully securing approval for TAT-7. This culminated in the liberalization of transatlantic communications in the mid-1980s. INTELSAT lost its monopoly over satellite communications and TAT-8, the first digital, fiber-optic, transatlantic cable tilted communications in favor of cables. New satellite technologies, such as London Teleport and SatStream, also shifted international communications away from domestic monopolies and toward those international service providers, among which BT could count itself, that could provide services circumventing domestic networks.

But this was more than a history of technology, of satellites and cables. The environments of these technologies were also key. In the 1950s and early 1960s, both AT&T and the Post Office presented themselves as conquering the oceanic and space environments on behalf of their nations. The Post Office used TAT-1 and Goonhilly to emphasize British technological mastery over the environment, while in the US, AT&T highlighted TAT-1 and Telstar's hazardous ocean and space environments to accentuate its contributions to the projection of US military and economic interests abroad. Once satellites began competing with cables, satellite organizations like INTELSAT and COMSAT wielded "one world" discourses and the environment as rhetorical devices to bludgeon cables, emphasizing cables' fragility and satellites' global and environmental transcendence. Meanwhile, the Post Office and AT&T worked on new technologies, from sea-ploughs to submersibles, to protect and maintain undersea cables. These strategies were not purely technological either, as new transnational corporate partnerships took form in the North Atlantic Cable Maintenance Agreement and the North Atlantic Systems Conference to defend and promote cables. Here, the environmental history of cable maintenance was an important foundation for the corporate partnerships that successfully opposed INTELSAT.

In the 1980s, however, the environment became much less visible in discourses around transatlantic communications. AT&T and BT, as newly denationalized telecom corporations, articulated a new information age discourse oriented to global capital. In this discourse, transatlantic communication was instantaneous and dematerialized. Satellites also became the preeminent symbol of global communications, no longer a communications technology but an "information technology," juxtaposed with computers to perpetuate the idea of satellites as the key transnational technology of the information age. AT&T and BT's advertising fused space age and information age discourses so that their international business clientele would see the world through the lens of a satellite.

This history also shows how new transatlantic communications technologies were used to build monopolies and markets in ways that complicate traditional narratives of the "market turn." Such a traditional narrative would point to the FCC's 1985 decision, during the Reagan administration, to liberalize satellite communications as a classic example of a monopoly, INTELSAT, being turned into a market. So too do AT&T's divestiture and BT's liberalization show how domestic deregulation meant that these corporations could compete more on international markets. Or, in the wider, global history of the market turn, broader international regulatory changes, like the General Agreement on Tariffs and Trade, especially the 1973–1979 Tokyo Round of negotiations, seemed to provide regulatory frameworks that facilitated the rise of neoliberalism.[98] But to focus only on these institutional arrangements would miss how, since the 1950s, the Post Office and AT&T had used submarine cables to disrupt the regulation of international communications. For the Post Office, TAT-1 was an opportunity to disrupt the FCC's separation of international telephony and telegraphy. During the 1970s, before their monopolies were deregulated, the Post Office, AT&T, and other North Atlantic telecom companies used their transnational corporate partnerships to fight INTELSAT's monopoly.

In short, this history shows that the Post Office, BT, and AT&T used the infrastructure of transatlantic communications to pressure for deregulation from the bottom up. It was infrastructural change, both in the form of new technologies, such as coaxial cables and communication satellites, and in the form of corporate partnerships, that mattered, rather than ideological change. In this light, liberalization and privatization were not ruptures but showed continuities on either side. The Post Office was increasingly focused on corporate markets beforehand and continued so as BT afterward. All throughout this history, the Post Office and BT's business customers played an influential role in motivating the search for cheaper transatlantic communications and deregulated international telecom markets. The next chapter focusses on those customers directly, showing how, through the 1970s, the Post Office became increasingly oriented toward London's financial sector in a relationship that would ultimately dictate BT's privatization.

7 THE CITY OF LONDON: FINANCING BT'S PRIVATIZATION

In March 1982, as part of Information Technology Year 1982 (IT-82), Patrick Jenkin, the Conservative government's secretary of state for industry, wrote to Leon Brittan, the chief secretary of the Treasury, to advocate BT's privatization, arguing that it could be "the most lasting legacy of Information Technology Year."[1] Nine months later, in December 1982, Margaret Thatcher opened a conference on information technology at the Barbican Centre, London, as part of IT-82. In her speech, Thatcher spoke about how information technology required free enterprise and competition, necessitating BT's privatization and liberalization. As well as BT, however, Thatcher also spoke about information technology's importance to the City of London, known simply as "the City", a metonym for the British financial services sector. Thatcher explained that information technology had "helped London to become the most efficient financial center in the world, through the City's ability to process vast amounts of information quickly and accurately."[2]

For Thatcher's Conservative government, information technology was not the only subject that connected BT and the City of London. Just under two years later, in July 1984, John Redwood, current backbench Conservative MP and then director of 10 Downing Street's Policy Unit, explained to Thatcher how the government's strategy of targeting the British public for shares of BT's sale would instead help the government sell BT to British financial institutions in the City of London.[3] The illusion that the government could sell large stakes of BT to other markets, such as the British public, would coax the City of London into investing larger sums in BT. In doing so, this would help avoid BT's sale failing, as the government worried it was too large to succeed. This strategy complicates the traditional narrative of BT's privatization as an act of "popular capitalism," as suggested by ads announcing BT's sale as "a public service goes public," telling audiences that they could become "an owner of a company."[4] This chapter argues that the ways that privatization and information technology policy connected BT and the City of London cannot be understood as separate from each other. The financialization of British telecommunications and the Conservative government's focus on information technology came together in BT's sale, which presented privatization as a necessary precondition for both the City of London's transformation into a global financial center and Britain's participation in the "information revolution."

This chapter is thus, in part, about how Conservative politicians understood digitalization. Much has already been said about the relationship between the Right and digitalization in the US. The US new Right appropriated digital utopian discourses about digital technology as individualized and emancipatory to advocate for deregulation and the "New Economy."[5] The quintessential example is the 1994 landmark essay "Cyberspace and the American Dream: A Magna Carta for the Knowledge Age," written by the journalist-investor Esther Dyson, the futurist Alvin Toffler, the economist George Gilder, and George Keyworth, a physicist who later became Ronald Reagan's chief science adviser and founded the Progress & Freedom Foundation think tank, which published "Cyberspace and the American Dream."[6] These authors described information technology as central to individual freedom and market power and were influential, particularly on Newt Gingrich. Gingrich, the Republican congressman and speaker of the US House of Representatives from 1995 to 1999, emphasized information technology as central to the deregulated "new economy" of the 1990s United States and was featured on the front cover of the digital utopian magazine *WIRED*.[7] By the 1990s, social, political, and economic ideas about digitalization had blended to produce a vision of IT as enabling individual freedom and entrepreneurialism, and shrinking the state through competition, deregulation, and the "electronic marketplace." But the influence of countercultural, digital utopian visions on the US right-wing does not explain the Thatcherist emphasis on information technology in BT's sale in 1980s Britain.

For that, this chapter turns to the City of London. The traditional narrative of BT's sale as an act of "popular capitalism" presents a particular narrative about Thatcherism and finance. In this narrative, privatization turned citizens into "financial consumers." Financial institutions and the Thatcher government promoted individual share ownership, for which BT's sale was an important test case.[8] Individual shareholders, however, rarely showed further interest in buying shares, and many quickly sold their stakes. Furthermore, as John Redwood's memo at the start of this chapter suggests, financial institutions, rather than citizens, were the Thatcher government's primary target for BT's sale. This sale was thus not a dramatic or enduring act of popular capitalism, and so this chapter instead investigate the deeper intersection of the Post Office and BT's history with the changes in Britain's financial sector in the 1970s and 1980s.

The single moment that dominates histories of British finance in this period is the City of London's 1986 Big Bang, when the Thatcher government deregulated finance and trading switched from the market floor to computer screens. The Big Bang supposedly transformed the City of London into an international financial center, shaping global finance, and was responsible, for example, for developing a modern global securities market.[9] Beyond the City, financial liberalization, as seen in events like the Big Bang, apparently drove the "Thatcher revolution" and the "neoliberal revolution."[10] In this narrative, market liberalization drove financialization, which in turn drove neoliberal policies such as privatization. But the Big Bang was more an "accident" than an "intentional revolution," an unanticipated consequence of deregulation agreements forged by the British government and the London Stock Exchange.[11] Beyond the Big Bang, the abolition of exchange controls in 1979 was also important, as was the longer postwar role of the Bank of England in promoting the City as an international financial center.[12] The City also campaigned for politicians to see finance as essential to Britain's economic growth throughout the 1970s, undermining social democratic policies intended to mobilize British finance in support of domestic industry. Instead, domestic finance managed to enroll British industry in its plans to financialize Britain, emblematic of a broader trend in organized business facilitating the rise of neoliberalism in the UK.[13]

This chapter thus investigates how the City of London enrolled the Post Office and BT by treating the financial sector as an influential, organized user group in British telecommunications. Studying users has long been an important focus of the history and sociology of technology, and users have been particularly influential in the history of telecommunications.[14] In North America, for example, early residential telephone users actively showed network providers that the telephone was not just an instrument for business but also for socializing.[15] In contrast, the telephone's social development in Britain was slower. High calling charges meant that, by the late 1930s, there were still few residential users, with the majority coming from the top 5 percent of income distribution.[16] Business, on the other hand was a much more active and influential user base for early British telephony. The commercial and brokerage sectors used telephones heavily, generally for routine information transfers.[17] After World War II, business users placed increasing demands on the public telephone system. As previous chapters have explored, data and international communications became more important to businesses during this period, especially financial institutions and multinational companies. At the time of BT's sale, it was clear that business users, especially the financial sector, had organized to lobby for privatization, and that this had influenced the Thatcher government.[18] These users believed that privatization would free BT of certain public service obligations, and so business users would not need to cross-subsidize residential users as heavily, and that a privatized BT could offer new services, better tailored to their demands. What is less clear is how these users shaped not just BT's privatization, but Conservative understandings of the relationship between information technology and finance, as well as the material infrastructure of British telecommunications.

Financial markets are, after all, not abstract informational networks. They are material, composed of people, objects, and tools.[19] Opening the black box of finance's materiality shows how technologies, such as high-speed fiber-optic links, digital screens, automated trading desks, and predictive pricing algorithms, have not simply eased and accelerated transactions but changed both the nature of markets and the way that people understand this nature.[20] This is by no means a product solely of digitalization either, as information and communication technologies, such as the telegraph and the stock ticker, have shaped global finance since the nineteenth century.[21] In the postwar period, managers and engineers computerized exchanges, from the New York Stock Exchange to the Chicago Mercantile Exchange, as part of strategies to consolidate control and maintain financial centralization.[22] The London Stock Exchange was also an enthusiastic adopter of new information and communication technologies.[23] From the 1960s and 1970s, "market engineers" captured British finance, aiming to move the LSE from diverse automated services projects onto single platforms, and found ways to make these infrastructures resilient to physical and digital threats.[24] These histories, however, tend to explore how the material infrastructure of markets has reshaped finance, rather than how finance might have shaped material infrastructures.[25] Furthermore, these histories tend to focus on how market engineers deployed technology within the financial sector rather than how these financial sectors connected to and relied on national and transnational infrastructures.

This chapter thus explores the City of London and the Conservatives' focus on BT and IT in two parts. The first part explores the role of financial interests in privatization and digitalization. The chapter shows how the City of London lobbied for telecom liberalization and privatization via its Telecommunications Committee, how financial interests influenced the design of BT's sale, and how BT came to focus on the City of London. In the second part, this chapter examines the relationship between BT's sale and Conservative information technology policy. Strengthening national "information" industries, such as finance, was a key issue, but information technology also fulfilled a Conservative ideological commitment to individual freedom and a small state. This chapter then highlights the broader influence of BT's sale on perceptions of the relationship between privatization and information technology, showing how BT's sale wedded the "information age" and the City of London.

FROM THE CITY OF LONDON TO "LONDON TELECITY"

In 1968, the Bank of England organized the City Telecommunications Subcommittee (CTC), a subcommittee of the Committee on Invisible Exports (CIE), which the Bank of England and the British National Export Council had created in 1966 to promote the City of London's "invisible exports," like financial services. The CIE was one of the City's key strategies for lobbying the government to privilege Britain's financial sector during the 1970s.[26] Leading figures from the City's financial community also filled its telecommunications subcommittee. It was chaired by Cyril Kleinwort, chairman of the Kleinwort Benson investment bank, which later became involved in privatization, managing the reprivatization of British Aerospace and flotation of Cable & Wireless in 1981, advising on the privatization of Associated British Ports in 1983, and managing BT's sale in 1984. Other members included W. M. Clarke and P. G. Vermeulen, director and secretary of the

CIE, respectively; R. van Koetsveld of the shipbrokers H. Clarkson & Son; and R. E. Liddiard, chairman of the British Federation of Commodity Associations. In 1969, Clarke took over as chair, and B. D. Townsend replaced Kleinwort as Kleinwort Benson's representative on the CTC.[27]

The CTC initially acted as the City's technical liaison to the Post Office. London's banks had been relying more and more on the Post Office's telecom infrastructure. The Bankers' Automated Clearing Services, which automated cheque and credit settlements between banks and was owned jointly by London's five largest clearing banks, launched in 1968, the same year that the CTC was created, and shared premises with the Post Office's computer center in Edgware, London.[28] Barclays had also already linked sixty branches over Post Office lines to two centralized EMIDEC 1100 computers since 1959 to maintain branch accounts and, from 1967, had networked two thousand branches with Burroughs TC500 computers, automating branch accounting, direct debiting, clearing operations, and more.[29] The CTC reported technical issues and requested updates on the rollout of services such as telex and data transmission. This relationship's cooperative dynamic shows in the minutes of joint CTC-Post Office meetings from the CTC's early years. Subjects discussed at these meetings include a brochure on overseas services that the Post Office's External Telecommunication Executive had created for London's businesses and offers from the CTC to track down fault locations and help manage the City's heavy use of telex.[30]

Even from this early stage, however, the CTC sought special treatment for the City. In the autumn of 1968, the CTC held a luncheon with the postmaster general, John Stonehouse, and requested a priority scheme for international telex service in the City of London. In December, the CTC received a reply from William Ryland, the Post Office's managing director for telecommunications, rejecting the CTC's proposal and explaining that, "The real difficulty is that it would not be realistic to expect that knowledge of a preferential service could be confined for any length of time to a very small number of customers. Once its existence became generally known, the service would, of course, be swamped and cease to have any value."[31] Here, Ryland rejected the telex priority scheme based on "uniformity," the principle that the Post Office could not offer one level of service to the city and a different level to everybody else. This would change after BT's creation and with further pressure from the City.

Further lobbying came in a 1973 report, "Present Shortcomings of International Telecommunications," by Walter Salomon, chairman of the merchant bank Rea Brothers.[32] Salomon complained that the City's users, "who most depend on high speed reliable communication across the international commercial and financial spectrum, are in danger of no longer being able to rely on this vital life line." He named poor operator service, delayed calls, outdated equipment, and too few international call routes as the City's four major grievances and, as a solution, argued for the privatization of the Post Office. This came during a period when international telecom networks were becoming more important for finance. Since 1971, sixty-eight banks in Europe and the US had been collaborating on the Society for Worldwide Interbank Financial Telecommunications, now known as SWIFT, a network that would turn international payment instructions from a mostly postal system to a system linked by international leased lines.[33] Salomon thus called for the Post Office to "be freed from State fetters, and allowed to develop through the media of private financial enterprise and international tender" so that it could prioritize these international connections for the City. Salomon's report was a call to arms, as he asked for the Confederation of British Industry (CBI), the British Bankers' Association, and the Accepting Houses Committee, a group of the City's leading accepting houses and merchant banks, to follow the CIE in exerting political pressure for the privatization of Britain's telecom infrastructure.

In response to Salomon's report, an ad hoc group, the City Telecommunications Group, formed from various organizations with City interests.[34] Representing the group's London-centric interest was chairman Geoffrey Finsberg, Conservative MP for Hampstead and, from 1974 to 1979, opposition spokesperson for Greater London. Alongside the CTC's original members and the organizations named in Salomon's report, representatives from the London Chamber of Commerce, the British Insurance Association, the Stock Exchange, and Lloyds of London also attended. The group discussed whether to pursue a cooperative approach with the Post Office or aggressively contact the press, publicize City dissatisfaction, and lobby the government. Clarke, the CIE and CTC's chair, had already obtained permission from Gordon Richardson, the new governor of the Bank of England, to raise the issue with Anthony Barber, the chancellor of the Exchequer, but a new report by the CBI, which took a similar line to Walter Salomon's report, divided members. Some were uncomfortable having their organizations' names associated with such a report while they still had to maintain good relations with the Post Office. The group thus agreed, to Salomon's discontent, that the CBI report would be toned down and that they would meet with the Post Office to establish a new direction.

The group met with Edward Fennessy, Ryland's successor as the Post Office's managing director for Telecommunications, in January 1974.[35] The group discussed the toned-down CBI report, and Fennessy's responses show how the telecommunications business began to change its attitude to uniformity and the City. Fennessy explained that the long lead time for increasing telephone capacity, and the anti-inflationary price restraints imposed by the government in 1972, meant that the Post Office found it challenging to address the City's complaints right away. He also noted, however, that the Post Office was investigating a possible twenty-four-hour fault maintenance service specifically for London's business houses. He concluded the meeting by saying that he "was particularly concerned to know the views of the City on development likely to be required over the next 10 years." The City Telecommunications Group and their calls for privatization had resulted in a small but significant shift from the Post Office's uniformity policy.

The Post Office did not have to wait long to hear the City's views. At a March 1974 meeting, tellingly titled "London as a World Financial Centre," Clarke, the CIE and CTC's director, and C. N. Read, director of the Inter-Bank Research Organisation (IBRO), a think tank set up by the City's banks in 1968, met with representatives from the Post Office's London telephone region.[36] Clarke and Read informed the Post Office that the CIE had, as a result of the City Telecommunications Group's concerns and its desire to maintain London as an international financial center, upgraded the City Telecommunications Subcommittee into a full committee, and invited the Post Office to participate. Clarke announced that the CTC's new remit was to deal with "the increasing inadequacies of international and domestic services" and that it would be composed of two joint City-Post Office working groups, one dedicated to the City's future telecommunication needs and one dedicated to current problems.[37] Trends from abroad may have influenced this focus on technology's importance to London's status as a world financial center. The New York Stock Exchange had led the way in computerization since the mid-1960s, and while Clarke did not mention it explicitly, the London Stock Exchange had sent a delegation to New York in 1970 to look at how the competition used technology.[38] Accordingly, the London Stock Exchange joined the expanded CTC, alongside the British Insurance Association, the Chamber of Shipping, and Lloyd's of London. From across the City, financial institutions had come together to align British telecommunications with their interests, mirroring the CIE as an avenue through which the City could lobby the government to restructure the British economy.

Despite the Post Office representation, the new CTC and its additional City membership became more aggressive in lobbying the Post Office. In 1977, Francis Sandilands, CIE chairman and CTC member, wrote directly to William Barlow, chairman of the Post Office, complaining about the Post Office's withdrawal of five-year fixed-price line rental contracts. Barlow replied that he appreciated the importance of telecommunications to the City but that the Price Commission had tied the Post Office's hands, ruling that the contracts contravened Price Code.[39] Barlow's reply thus accepted that the City might need special treatment, suggesting the Post Office's amenability to City interests, and that it was public ownership that prevented a special relationship between finance and telecom. This may stem from Barlow's own background and ambitions. Barlow had been appointed Post Office chairman in 1977 under the impression that telecoms would split from the Post Office and that he would lead the new telecommunications corporation with greater freedom from government. This was stymied, however, by the industrial democracy experiment covered in earlier chapters, which Barlow resented. In 1979, when Keith Joseph ended industrial democracy and announced that the government would create BT and explore ending its monopoly, Barlow immediately called for privatization.[40] Barlow thus matched the City in recognizing its claim to special treatment and joining its calls for privatization.

After Joseph's announcement, the CTC lobbied further for marketization and privatization. A 1979 CTC paper by the Foreign Exchange and Currency Deposit Brokers' Association shows the range of issues on which the City campaigned.[41] These included the liberalization of customer premises equipment, so that businesses could buy terminals such as telephones and telex machines from third-party suppliers; a new regulatory authority to oversee the Post Office and its successor, BT; and the complete liberalization of the telephone network, allowing new networks to compete. The paper also attacked uniformity, arguing that the "Post Office interprets this, when it suits them, as meaning that they may only offer equipment in the City of London that they will also offer to a crofter in the Outer Isles." Joseph's announcement also spurred the formation of new City-influenced lobbying groups. In 1979, Barclays, alongside Citibank, Sainsbury's, Smedley-HP Foods, and Blackwell's, founded the Association of Telecommunications Users, which lobbied for privatization and liberalization.[42] Barclays is noteworthy as it was one of the companies, alongside BP and Cable & Wireless, behind the creation of Mercury, the new telecom operator that competed with BT in a duopoly until 1990. In 1980, the British Bankers' Association also produced a report, "Telecommunications in the City," which again attacked the uniformity principle and asked BT, now that it had been set up as an operational unit in advance of its formal creation by the 1981 British Telecommunications Act, to review this principle urgently.[43] These calls all reinforce that the City saw liberalization and privatization as means to secure favorable telecommunications services for financial institutions.

City lobbying thus came along two lines. First, that the government should end BT's monopoly and privatize it and, second, that BT should end its uniformity principle. The City's success is evident in two banks' roles in designing BT's privatization and BT's new focus on the City. Two City institutions, Kleinwort Benson and Barclays Merchant Bank, played significant roles in shaping BT's privatization as an apparent act of popular capitalism and orchestrating the sale of nearly half BT's floated stock to City financial institutions. The government had appointed Kleinwort Benson to manage BT's privatization. Kleinwort Benson had advised and managed the sales of British Aerospace, Cable & Wireless, and Associated British Ports and, as noted above, its chairman, Cyril Kleinwort, was also the founding chair of the CTC. Kleinwort Benson valued BT at more than £3 billion, and both the government and the City of London worried that Britain's capital market could not bear so large a flotation.[44] In February 1984, a report for the Cabinet by Barclays Merchant Bank thus proposed targeting individual investors to raise capital and suggested that a beneficial side-effect of this strategy would be developing a "share-owning democracy" in Britain.[45] The report also noted that the City's traditionalism had created an aversion to individual share-ownership, and so targeting individual share-owners would create a new market to compete with the City for BT shares. John Redwood, in his note to Margaret Thatcher on this subject, reinforced this point, arguing that the key to getting City institutions to invest was creating the illusion that the government could sell substantial stakes in BT to other investors and so, because selling huge stakes to overseas financial institutions was politically problematic, individual investors became key.[46]

The government thus targeted the British public as another source of investment, partly to find extra capital and partly to create an illusion of competition for City financial institutions. In 1984, there were only 1.8 million individual investors in Britain, and forecasts suggested that the government would need to raise £1 billion from individual applications. A coordinated PR campaign began, with advertising on the TV, radio, posters, and in the press.[47] A special train promoting the share issue visited seventeen cities from May 1984, while BT's senior management put on a traveling roadshow that toured Britain, Paris, Frankfurt, and Amsterdam. The campaign aimed to show BT as a high-tech institution worthy of investment and convince the public that privatization was a way of giving back. From 1984, ads with the taglines, "You can share in BT's future" and "A public service goes public" appeared.[48] By early September 1984, the BT Share Information Office in Bristol had received more than 300,000 requests for information and Market and Opinion Research International, the market research company, reported that 630,000 citizens were sure to buy, with another 3.9 million likely.[49]

At 8 a.m. on Impact Day, Friday, November 16, 1984, underwriters in the City began to accept applications for BT's 3,012 million shares, valued at 130p per share. Within three hours, they had underwritten 2,600 million of those shares. That morning, Geoffrey Pattie, Kenneth Baker's successor as minister for information technology, announced in the House of Commons that Kleinwort Benson was holding promising discussions with City institutions to purchase around half of BT's offered shares.[50] Twelve days later, at 10 a.m. on Wednesday, November 28, shareholder applications closed. The public had submitted two million applications, outnumbering the number of individual shareholders in the entire nation before BT's sale, and the flotation was hugely oversubscribed. Total sale proceeds reached £3.863 billion and, less the costs of shareholder incentives, marketing, and underwriting and advising fees, came to £3.6 billion.[51] More than six times larger than the government's previous sale of BP at £566 million in September 1983, it was, at that time, the largest stock flotation in world history.[52] Kenneth Baker, the former minister for information technology, later wrote that BT's sale "made possible all other public utility sales," while Margaret Thatcher, in a 1992 speech, called privatization "one of Britain's most successful exports," trebling individual shareholders and stopping inefficient management.[53] In the end, however, Kleinwort Benson sold 47 percent of BT's shares to the City, 39 percent to individual investors, and 14 percent overseas.[54] Having lobbied for BT's privatization and shaped the sale strategy to entice more financial institutions, the City had successfully arranged to transfer most of BT's offered shares not from state ownership to citizen ownership, but to City of London financial institutions.

In another City victory, BT also ended its uniformity principle. This started in October 1980, shortly after BT came into being, when Alex Reid, BT's director of business systems, responded to pressure from groups like the CTC and the British Bankers' Association with London TeleCity.[55] London TeleCity meant that the City would be first to receive new services and infrastructure from BT, which Reid claimed would benefit the City, BT, and "the National Economy." The City was thus the first to receive X-Stream and ISDN services (discussed in chapter 4), and, in October 1983, was one of the first areas in the UK to get high-speed fiber-optic communication cables.[56] BT's press release proudly announced that London was one of the world's first cities to get fiber-optic and highlighted the business services that the City of London would now be able to use. BT also emphasized that these fiber-optic links would connect the City to London Docklands, where BT was building London Teleport, another piece of dedicated City infrastructure. London Teleport, Britain's new satellite earth station and its third after Goonhilly and Madley, was thus installed in London specifically to provide the City financial district with better international communication links. London Teleport opened on February 1, 1984, and used the City's new fiber-optic network to bring "high-speed telecommunications to the fingertips of the City."[57] By 1985, two crucial services for British finance, the Bankers' Automated Clearing Services, and the Clearing Houses' Automated Payments System, were also networked over BT's SwitchStream packet-switched network.[58] Through London TeleCity's rollout from 1980, the year of BT's creation, to 1984, the year that the City became BT's second-largest owner behind the government, the City received digital data services, fiber-optic cables, and a dedicated satellite earth station before any other customer group. As Alex Reid openly acknowledged, the pressure from the City meant that BT had ended its uniformity principle and replaced it with a system where BT prioritized the City above all others.[59]

This history casts a new light on BT's privatization, showing City organizations pressuring for the end of uniformity, lobbying for BT's privatization, and co-opting the "share-owning democracy" ideology to increase BT's appeal to financial investors. These findings, however, should be placed in context. Barclays Merchant Bank did not coin wider share-ownership *de novo* in 1984 to increase BT's appeal. Wider share-ownership had been an ideological commitment inside and outside the Conservative Party since the late 1950s. In the mid-1970s, with Thatcher's arrival as party leader, Conservative politicians, such as Keith Joseph, began to see wider share-ownership as a method for promoting individual freedom.[60] Furthermore, another important motivation for BT's sale was to remove it from the public-sector borrowing requirement, freeing BT to borrow additional capital to invest in the network. As noted several times throughout this book, government spending restrictions had constrained BT throughout the 1970s, and by 1982, George Jefferson, BT's chairman, was pressuring Patrick Jenkin to privatize so that BT could escape the PSBR.[61] Finally, it is likely that, in any event, financial institutions would have necessarily been a large part of any sale, as BT's £4 billion privatization was too large for individual shareholders alone. The City of London, however, was not just part of these events, but actively shaped both BT's sale and its direction of development. Furthermore, BT's sale played a further important role in linking privatization, the City of London, and the "information revolution."

PRIVATIZING THE INFORMATION REVOLUTION

Information technology occupied a unique position within Thatcher's industrial policy. In 1981, she created a post of minister for information technology, the first of which was Kenneth Baker, nicknamed "Mr. Chips" in the press after the government's fixation on microchip manufacturing.[62] As previously discussed, Thatcher also announced 1982 as IT-82, a National Information Technology Year, a decision that legitimated "information age" narratives worldwide.[63] Patrick Jenkin, Thatcher's secretary of state for industry, in 1982 also suggested that BT's privatization could be the "most lasting legacy" of IT-82.[64] When Kleinwort Benson's sale of BT was announced to the House of Commons, it was by Geoffrey Pattie, Baker's successor as minister for information technology, rather than Norman Tebbit, the secretary of state for trade and industry. Why did the Thatcher government emphasize information technology policy, and why was BT's privatization central to this policy focus?

The broad strokes of an answer appear early in the Thatcher government's first term. In 1979, a Department of Industry report on liberalizing British telecommunications argued that public monopoly would continue to "weaken London's strength as an international centre of commerce," and proposed a new network, owned by a consortium of clearing banks, BP, Cable & Wireless, and ICI, the chemical company.[65] In July 1980, Keith Joseph subsequently announced that the government would liberalize telecommunications in Britain, ending BT's monopoly and, in 1981, approved the creation of Mercury, a new network owned by Cable & Wireless, BP, and Barclays, because of the "dreadful service" that BT was providing the City.[66] From this early stage, prioritizing the City of London was a clear objective for information technology policy. But, in addition to focusing on communications, the government also set up the Alvey Programme, a "strategic computing initiative" to sponsor information technology research.[67] The Alvey Programme was named after John Alvey, senior director of technology at BT, who chaired a report responding to Japan's perceived information age trailblazing, notably its Fifth Generation Computing Project. Projects like Alvey and FGCS, as well as the Strategic Computing Initiative in the US and ESPRIT in Europe, all show governmental privileging of information technology in the early 1980s but represented different political economies of digitalization. SCI was the product of a militarist-capitalist political economy, FGCS a state-led socially responsible political economy, and ESPRIT a European liberal-integrationist political economy.[68] Alvey, meanwhile became acceptable to Thatcher as a large injection of public funds into the private sector because it was framed as near-market industrial research, which would potentially help entrepreneurs and innovators more than large corporations.[69] Early IT policy under Thatcher thus seemed to focus on privatizing communications to support finance, and admitted public investment only when framed as a near-market exercise.

Deregulating BT was about more than just supporting finance, however. It was also about turning BT from a national into an international corporation. Patrick Jenkin had informed Parliament in 1982 that one of the main goals of privatization was to turn BT into "a major world force."[70] This manifested in the Thatcher government's seemingly contradictory policy toward digital exchange procurement, first protecting and centralizing, then opening to foreign supply (as discussed in chapter 3). This policy was in fact about helping BT upgrade its network, which BT's financial users thought was sorely in need of improvement. BT itself took action to compete more on international markets in this period, securing more business users, and this was supported by the Thatcher government through liberalization and privatization (as discussed in chapter 6). A 1982 white paper announcing the government's plans to privatize BT, *The Future of Telecommunications in Britain*, explained how privatization meant that the government would replace BT's public-sector borrowing requirement with IT regulations that would be "the most liberal in the world" and so "would free BT from traditional forms of government control," allowing it to compete internationally.[71] The paper further equated information technology with market power, explaining that "competition and the advent of new technology are stimulating BT to respond to market opportunities." Privatization was about giving BT more opportunities to serve its business customers, rather than providing competition to make BT work harder for its residential users. This is further shown by how the Thatcher government fought amendments introduced in the House of Lords that would promote competition for BT after it was privatized.[72] So, while the City was not opposed to liberalization, it was more invested in privatization as an act that would orient BT toward the financial sector. Supporting finance and enhancing BT's international reach were thus important components of the government's information technology policy.

But IT policy, and especially BT's privatization, also provided the Thatcher government with a vehicle to articulate ideological commitments to individualism, entrepreneurialism, market power, and a small state.[73] A speech, "Towards an Information Economy," by Kenneth Baker during IT-82 demonstrated information technology's ideological appeal to Conservative ministers.[74] Speaking to the British Association for the Advancement of Science, Baker contrasted the opportunities of information technology in the "post-industrial society" for greater personal freedom, the retreat of the state, and privatization with a dystopic vision of the "Electronic State." The electronic state had the power "to survey, control and manipulate the citizen," which Baker compared to the technologically manipulated societies of Aldous Huxley's *Brave New World* and George Orwell's *Nineteen Eighty-Four*. Baker did not explicitly reference BT's impending sale but portrayed privatization as a weapon against the electronic state, arguing, "We should enhance the opportunities of private ownership for what the State owns it has to control. The State will provide much, the Electronic State could provide more, but it would exact a price in terms of personal freedom." For Baker, privatization was an essential precondition to realizing IT's individualist, emancipatory, free-market potential and preventing its Orwellian applications.

Baker's speech shows the politics of privatization as an alternative site of the digital utopian ideologies appearing in the US from the 1970s.[75] Baker called the "free flow of information" a necessary condition for a liberated information society, greater personal freedoms and private ownership, and the retreat of the state.[76] This sentiment resembles, yet predates, WIRED founder and cyberculture guru Stewart Brand's maxim that "information wants to be free," which has since become one of the rallying calls of digital utopianism.[77] The "free flow of information," however, has a longer history as a technical term from economic and industrial policy in the late 1970s and early 1980s. The "free flow of information doctrine" was a policy program supported by the US and British governments, the OECD, and transnational companies such as Coca-Cola and IBM.[78] It meant an openness to "transborder data flows," which were necessary for many transnational companies, especially financial institutions, to operate across multiple countries. Opposing the free flow of information thus meant opposing global capitalism, and so deregulating and privatizing telecommunications was cast as essential to promoting trans-border data flows. Baker's speech shows the early association of deregulatory information technology policy with free-market and individualistic ideologies, brought together in IT-82 and the privatization of BT.

The government also targeted BT's management with this discourse, presenting information technology as a necessary part of economic liberalization, which would release BT to expand commercially. In a July 1982 Q&A with senior BT managers, Patrick Jenkin presented privatization as necessary for the information revolution, explaining that Britain "cannot afford to keep BT trammelled by the mesh of bureaucratic controls at a time when technological and commercial developments really set this organisation at the centre of our electronic future."[79] BT included this message in its internal communications campaign to persuade staff about privatization. As the 1980 Long Range Strategy Seminar (discussed in chapter 2) showed, these readings of information technology were popular among BT's senior management, who understood liberalization as necessary to BT's technological development, which in turn promised to create a small, individualistic state. BT's senior management was thus quite amenable to privatization. In April 1982, George Jefferson, BT's chairman, wrote to Patrick Jenkin to explain that he was keen to explore any options that would remove BT from the public-sector borrowing requirement constraints, despite Post Office Engineering Union resistance.[80] That said, BT's senior management was concerned about the strain of privatization on staff, also reminding Jenkin that the "ingrained attitudes" of the Civil Service among staff would "take a long time to change."[81]

BT's management deployed popular capitalist narratives to convince their staff about privatization. Employee share-ownership formed a significant part of BT's internal strategy for involving staff in privatization. Drawing inspiration from the growth of employee shareholding in early privatizations of Britoil, British Aerospace, and Amersham International, the potential for employee shareholding in BT had been identified by November 1982, and by January 1983 was seen as a "highly desirable" strategy for involving staff and bringing the unions onside.[82] Kenneth Baker and Cecil Parkinson (secretary of state for trade and industry) were also influential supporters, meeting with George Jefferson in June 1983 to discuss BT employee share-ownership.[83] Baker was such a staunch supporter that, later in 1983, he inquired if BT could offer staff £600 loans to buy even more shares during the issue, to which Jefferson responded with appreciation for Baker's enthusiasm, but declined because of the legal and tax implications, as well as the moral argument that they should not encourage staff to get into debt.[84] Regardless, BT still set a UK record for employee share-ownership, with 10 percent of shares reserved for BT staff and pensioners.[85] BT's scheme was considered a great success, with 96 percent of staff applying for shares, ignoring union directives.

BT and the Conservatives thus both combined popular capitalism and a new discourse of digital liberalism to promote BT's sale. But where popular capitalism's origins in the drive to attract financial institutions to BT's sale is clear, the origins of the Conservative government's discourse of information technology as a technology of individual freedom and small government is less so. Neither Conservative politicians nor BT executives referenced specific figures or works that inspired this discourse, but Kenneth Baker's mention of the "post-industrial society" in his speech "Towards an Information Economy" suggests the influence of Daniel Bell's 1973 book *The Coming of Post-Industrial Society: A Venture in Social Forecasting*.[86] Bell's ideas had influenced the Post Office's long-range planners (as explored in chapter 2), but Bell's work was also influential politically, helping politicians understand the increasing presence of computers in society during the 1970s.[87] By reading mental freedom—"free thinking"—as essential to knowledge production, various US politicians and entrepreneurs began to believe that, as information technology enabled knowledge production, so it would by extension enable individual freedom.[88] In the late 1970s and early 1980s, the microcomputing industry in Britain took off, and so, as in the US, theories of the postindustrial society and personal computing fused with ideological commitments to individualism, producing visions of information technology as a technology of individual freedom that necessitated the shrinking of the regulatory state.

Even left-wing politicians agreed on information technology's liberating qualities and BT's importance to the information revolution. In a 1982 film, *New Technology, Whose Progress?*, Tony Benn attacked IT-82 as promoting the use of IT "to remove decision-making from the worker and increase management control."[89] Benn, however, still saw information technology as emancipatory, saying that it could "give people a sense of freedom" and instead argued against the entrenchment of corporate power through IT. In 1984, when Geoffrey Pattie announced Kleinwort Benson's sale of BT to the House of Commons, Alan Williams, the Labour MP for Swansea West, complained that the government "intend to hand over to the whims of short-term profit maximization the very industry that will be at the center of the information technology revolution."[90] Williams' response showed that left-wing politicians, too, believed in the promise of the information revolution and BT's value to that revolution, albeit under state ownership.

BT became important abroad as well, as its sale cemented the importance of privatization to information technology policy. Tom Forester's 1987 popular science book *The High-Tech Society* referred to the privatization of BT as a necessary move to market access, required to realign the telecommunications industry for the IT revolution.[91] This became the European consensus, with the liberalization and privatization of BT serving as "the model for European telecommunications deregulation" on the way to the European Commission's 1987 Green Paper, which advocated telecom deregulation.[92] This was followed by a 1994 European Commission report, "Europe and the Global Information Society," by Martin Bangemann, the commissioner for the internal market and industrial affairs under Jacques Delors, which further advocated privatization as a necessary condition for spreading the information revolution, a "market-driven revolution," throughout Europe.[93] As Thatcher had said in her 1992 "Principles of Thatcherism" speeches, privatization had indeed become one of Britain's most successful exports, but what she overlooked was how, with BT, one of the earliest and largest telecom privatizations, Britain had become the leading example of how a country should kickstart its information revolution.

BT's privatization also provided a chance to explicitly connect the City of London to the information revolution, so casting the financial sector as an information industry. Margaret Thatcher, in her speech on information technology at the Barbican in 1982, positioned IT as underpinning both BT's privatization and the success of the City of London. This came after a Central Policy Review Staff paper on "information technology" in 1980 had already described how the "communications revolution" had radical potential to transform the financial sector, and specifically highlighted Britain's telecom infrastructure as crucial to this revolution.[94] Banks had joined politicians in emphasizing this importance as well. When Gordan Richardson, governor of the Bank of England, opened National Westminster Bank's new Management Services Centre in 1979, he described its "massive battery of computers and automated equipment" as "not just a collection of silicon chips and transistors, but the means of keeping British banking in the forefront of a very competitive league."[95]

In the most public reversal of its uniformity policy, BT also proclaimed its new attention to the City during privatization. During 1984, BT ran an advertising campaign, "The Power Behind the Button," in which it promoted privatization as essential to unshackling BT's technological sophistication and power.[96] Ads in the first phase showcased BT's wide range of technological products, from optical fiber to System X to Goonhilly earth station. In the second phase, ads focused on the domestic setting of the home telephone, addressing anxieties about privatization and customer service. The third and final phase returned to high technology and emphasized BT's participation in the privatization as essential to the success of the City of London. A TV ad from this phase highlighted services such as BT's City Business System, which money dealers used to place global telephone calls and transactions, and closed with "Helping London stay at the heart of the world's financial markets, British Telecom is the power behind the button."[97] Thatcher's speech and BT's ad portrayed the City of London, alongside electronics manufacturing, as another national industry important to Britain's information revolution. Moreover, as government ministers repeatedly cast BT's sale as essential to Britain's participation in the information revolution, the implicit message was that privatization made Britain's information revolution and the City of London's financial success codependent.

In the privatization of BT, neoliberal ideologies about information technology thus met with the financialization of telecommunications infrastructure. BT had become central to narratives about an entrepreneurial, emancipatory information revolution, in which BT made the UK's information revolution happen by focusing on the City of London. This contrasted with 1968, when the CTC formed and when the Post Office would not break with its uniformity principle, nor did politicians position the Post Office or information technology as central to individual freedom and the small state. These changes came from many directions. Institutions from the City of London lobbied for BT's privatization, ensured that financial interests became BT's primary purchasers, and caused BT to prioritize the City of London over the rest of the country. Meanwhile, in contrast to early centralizing approaches to computing, political ideas about information technology's importance to the national economy resulted in a politics that prized deregulated communications and free-market individualism. For the Conservatives, liberalizing and selling BT was key to achieving both, providing an opportunity to solidify BT and the City of London as global players while demonstrating an ideological commitment to individualism, free markets, and the small state. Privatizing telecommunications thus became a necessary precondition for all three, of Thatcherism, financialization, and the information age. One of the earliest and largest privatizations, BT's sale produced the broad consensus that privatizing telecommunications monopolies was the fastest route to the information revolution. This peaked when Margaret Thatcher and BT fused privatization as an act of financialization and digitalization, presenting the City of London as a high-tech industry that privatization and the information revolution would make world-leading.

CONCLUSION: THE LONDON IDEOLOGY

This chapter opened with the premise that BT's privatization as an apparent act of "popular capitalism" cannot be understood as separate from Conservative information technology policy that emphasized both BT and the City of London. While the government's popular capitalism strategy at first seemed a method to increase BT's appeal to the existing market in the City of London, it is clear that a longer history of the City's relationship with the telecommunications business shaped this strategy. Since the early 1970s, City institutions had called for privatization so that the telecommunications business might start favoring the City over the rest of the country. The City felt this need so urgently that it formed various pressure groups, from the City Telecommunications Subcommittee to the City Telecommunications Group to the Association of Telecommunications Users. These groups successfully pressured for BT's sale, realizing the City's long ambition to get preferred customer status from BT, showing the vital role of business and financial interests in shaping Thatcherism and popular capitalism.[98] BT's sale kindled the share-owning democracy, but did not ignite it, and clearly, for the government and BT, the more important customer base lay in the City of London. This customer base's power was twofold, deriving from its status as both users and investors. The City could pressure the Post Office and BT not only because it was a well-organized user group but also because it represented the largest group of potential investors in BT's sale. Beyond this, however, BT's sale also tied financial interests to the Thatcher government's politics of information technology, and not just its politics of privatization.

Exploring the City's role thus also addressed this chapter's second task of investigating how BT's privatization influenced information technology policy and rhetoric. Politicians and BT executives aligned privatization with the "information revolution" by arguing that information technology realized their ideological values of individualism, market power, and a small state. Beyond this, however, were two revealing moments from Thatcher and BT, in which Thatcher's IT-82 speech and BT's "The Power Beyond the Button" advertising campaign showed that prioritizing the City of London was an essential part of the British "information revolution," and that privatization would enable this. This discourse shows some similarities to the digital utopianism and the Californian ideology of the 1990s in the US, in which information technology provided the place for countercultural values of individual freedom to meet with deregulatory economic policy.[99] The events described in this chapter, however, predate 1990s America and illustrate how political and financial interests shaped this ideology.

BT's privatization was thus central not to a Californian ideology but to a "London ideology." The London ideology drew on existing framings of information technology as emancipatory, but then cast the privatization of information technology as essential not only to realizing this emancipation but also to shrinking the state, facilitating free markets, and empowering

the City of London. The privatization of telecommunications became interpreted as a prerequisite for the information age worldwide, from popular
writing in Tom Forester's *The High-Tech Society* to policymaking in Martin
Bangemann's report, "Europe and the Global Information Society." The London ideology has influenced both the privatization and financialization of
digital communications infrastructure around the world. Thatcher called
privatization, as an act of industrial efficiency and popular capitalism, one
of Britain's "most successful exports." Since then, popular capitalism has
disappeared but the London ideology, fueled by BT's sale, remains. It was
Thatcherism's most successful export.

CONCLUSION

This book began with what was once the world's largest stock flotation, the 1984 privatization of British Telecom. While BT's privatization was central to Thatcherism and popularized utility sales within and outside the UK, this book has shifted the focus from that single moment to ask, more broadly, how national ownership and privatization mattered to digitalization, and vice versa. The history of Britain's telecom infrastructure, from the middle of the 1950s to the early 1990s, captures this intersection in a way that no other example can. This book argues that these processes were central to each other. Nationalization and privatization shaped digitalization, and this is perhaps not a surprising finding. It follows intuitive understandings of the relationship between politics and technology. Margaret Thatcher's Conservative government privatized BT in 1984, and this changed the trajectory of Britain's digital communications infrastructure from public to private, monopoly to market. But digitalization also shaped national and private ownership, and that is this book's more important finding. Digitalization constrained what was and was not possible and, in doing so, shaped how managers, engineers, and politicians enacted nationalization and privatization.

Britain led the way in privatization during the 1980s. The Thatcher government began with the oft-forgotten council house sales and continued by selling smaller state-owned enterprises such as British Aerospace and Cable & Wireless. Between 1979 and 1984, the government sold its stake in sixteen enterprises that were either wholly or partly state-owned. None of these sales breached £1 billion until November 1984, when BT sold for £3.6 billion.[1] But when the government sold BT, privatization was not a uniform policy that the government understood and applied in the same way for all state-owned enterprises. As chapter 7 showed, Conservative politicians recognized what scholars have not: that privatization was not general, but specific, formed in dialogue with the material and technological nature of the industries and infrastructures being sold. BT's privatization was not just the sale of a state-owned enterprise, but the sale of a digital infrastructure, and so it shaped and was shaped by technological change, not just by political and economic change.

In BT's case, its privatization was as much a part of digitalization as digitalization was part of privatization. Understanding one requires understanding the other, so this book turned to the history of digitalization in Britain's telecom infrastructure. This book's first key finding was the importance of visions and plans under national ownership to digitalization. In the early 1960s, after the analogue failure of Highgate Wood, engineers eyed digitalization's potential. They created a totalizing vision, influenced by cybernetics and Britain's "government machine," that encompassed voice, video, and data and united information with administration. These were not the only influences. The threat of competition from special-purpose, packet-switched data networks made this vision defensive and monopolizing, and the Post Office's corporatization in 1969 fueled technocratic management. Corporatization also led to a new long-range planning department that sustained and mutated this vision. Post Office futurologists developed computer simulations that chose technologies that supported their digital vision, and they found and spread new ideas about the digital age. Computer simulation also gave new ways to manage liberalization and privatization, imparting new understandings of the corporation and the market. Finally, and perhaps most importantly, futurology helped managers and engineers form new digital visions. With BT's futurology department and its predictive simulations, new visions emerged of the power that digitalization could give over customers and the marketplace.

As the Post Office's projects showed, however, infrastructures do not easily follow master plans such as these.[2] The Post Office tried to implement its digital vision through computer control and high-bandwidth integrated transmission. Computer control spread throughout administration to influence both switching and transmission. Simulations backed TXE4 as the infrastructure's intermediate electronic exchange, paving the way for the all-digital System X, and backed the waveguide as Britain's high-bandwidth backbone. But these events showed both the limits and reach of these plans. The Post Office and BT failed more than succeeded in developing for a high-bandwidth, integrated digital infrastructure, in which they would have a monopoly over both television and telecom distribution. The ISDN standard justified an austere approach to public investment in digital infrastructure. Digital integration even gave prospects for competition, offering a way for cable TV providers to compete as telecom companies. Integration, born to shield monopoly, became the spear of the market. On the other hand, digitalization supported the Post Office and BT's monopsonies over suppliers and labor. Automating long-distance dialing with GRACE meant that the Post Office imposed new standards of "friendliness" on its telephone operators. The TXE4 simulation, which first caused a national controversy, ended with the Heath government reaffirming the Post Office's power over its suppliers. System X, launching during liberalization, put technicians and operators out of jobs, and changed the nature of labor for those that remained. It also offered a way to reorganize Britain's electronics industry in favor of BT, even after denationalization. Switching thus shows how digitalization preserved and even extended the telecom business's power across privatization and liberalization, while transmission shows how privatization and liberalization inverted engineers' vision of digital integration.

Computerization and transmission were not the only projects that showed the Post Office and BT's digital vision. One project that this book did not explore in detail was Viewdata. Developed by the Egyptian-born Post Office engineer Samuel Fedida in the early 1970s, Viewdata built on Viewphone to offer interactive information services through customers' television sets.[3] Viewdata was an early example of "videotex" systems, a family of two-way interactive media systems developed from the 1970s that converged telephone and television systems. Fedida had an experimental system running by 1972, and Post Office management was very keen, perhaps because Viewdata built on Viewphone, already an icon of the Post Office's integrated digital approach. Viewdata launched in 1979 with a new brand name, Prestel, supplying 100,000 pages of online information to customers. Prestel, however, was a failure. It required an expensive, specialized Prestel TV set, and by the end of 1980, it had only six thousand users. By 1982, this had tripled to eighteen thousand users, but only twenty-five hundred were home users. Rather than bring the digital vision to the home, Prestel was instead most popular with business users, such as travel agents. Prestel proved particularly inspirational to one of BT's most valued user bases, the financial sector in the City of London. At the London Stock Exchange, engineers built on Prestel to create TOPIC, a "super Prestel" that gave a way to display market information and react to it from a distance. TOPIC "opened the floodgates of market information," giving brokers and jobbers a way to move the trading floor away from physical interactions, and so virtualized the marketplace.[4] Despite Prestel's committed business users, it was still considered an expensive failure, peaking with ninety thousand users. When BT turned Prestel off in 1991, this number had declined to only twelve thousand users.

Prestel has been called the "most extensive, high profile, and most ambitious online service available to home users in 1980s Britain."[5] Its failure might thus suggest disinterest in online communication among British home computer users, who were more interested in educational computing and video games. But Minitel's success in France shows a huge appetite among the public for online services, so Prestel's failure is perhaps better understood through comparison with Minitel.[6] Prestel was expensive and its rollout unsubsidized, whereas the French government supplied every home with a Minitel terminal in place of their ordinary telephone directory. The Post Office and BT played a strong gatekeeping role with Prestel's databanks, hosting all the content on a centralized computer in London called DUKE, whereas Minitel's network was open to private vendors to host data, giving them more autonomy. Prestel's failure thus seems to indicate some of the limits to the Post Office's centralizing, monopolistic, integrated digital vision.

This vision was not the only way that the Post Office and BT worked through digitalization. Digitalization, nationalization, and privatization had various places and spaces, and this book has shown that these places' histories gave greater insight into the local, national, and international histories that co-constructed infrastructure ownership and technological change. The Post Office and BT played a central role in making a new place, Martlesham Heath, which contained an IT park, one of the digital age's quintessential spatial formations. Martlesham Heath began as a social democratic project as government dispersed its work and published the "South-East Study" to develop the region. Martlesham Heath soon privatized and denationalized, however, as the private sector built a new village that evoked English history, while Adastral Park became BT's home node for its international, collaborative corporate ventures. One of the Post Office and BT's key partners through this history was AT&T, and the North Atlantic was central to this partnership.

Through North Atlantic communications projects, AT&T, the Post Office, and BT worked to disrupt the regulation of international communications so that they could carve out a bigger share of this growing market. These international markets were of such particular concern because of their importance to one of the Post Office and BT's most active user bases, the financial sector in the City of London. Through a decade-long lobbying campaign, the City of London's financial sector convinced BT to reorient its digitalization efforts from the nation to the City of London. This forged a political view that the City was, like BT, an information industry and that liberating one, by privatizing BT, would liberate the other, fast-tracking the British economy into a financialized information age. This view, which stretched across engineers, managers, financiers, and politicians, showed the convergence of digitalization with privatization as an act of denationalization and neoliberalism. In doing so, Britain built a new political economy of telecommunications.

One of this book's key goals was to analyze how infrastructural change affects major transitions in political economy, by looking at how digitalization in British telecommunications influenced the "market turn" from a nationalist to a neoliberal political economy. The brief life of British social democracy should not be forgotten here. In the 1970s, declining economic nationalism gave room for social democracy to breathe, and, from finance to urban planning, a range of genuine social democratic alternatives appeared.[7] But it would be a mistake to assume that postwar experts and technocrats were social democrats, and that holds true for telecommunications.[8] To be sure, there was a public interest in their plans to build a digital infrastructure serving all forms of information to every home and business, and the principle of uniform service motivated senior management into the 1970s. But uniformity ended, and this vision was only ever partially realized. Labour and Conservative governments alike never pursued the Post Office and BT's offers to build a nationwide infrastructure, via coaxial cables or optical fiber, that would provide voice, video, and data.

Rather than promoting a social democratic vision of digitalization, engineers were most successful in their technocratic, corporate agenda of computerizing communication and administration. Their projects automated labor, computerized management, and distanced users. By the 1980 long-range strategy seminar, computerization offered a new vision of BT's digital domination of customers and the marketplace. Even the vision of an integrated digital network was never particularly welfarist. Instead, it was a vision, fueled by bureaucratic, cybernetic ways of thinking about administration, that was advanced to defend the Post Office's monopoly, and was entrenched by the Post Office's corporatization. For the Post Office's engineers, the transmission of information was, as Merriman told the Post Office board, simply "PO business."[9] This motivation stands in contrast to the nationalist and social democratic goals of Highgate Wood and the Post Office Tower, the last two major switching and transmission projects before the digital vision took shape. The nationalist political economy of digitalization was, in fairness, quite resilient, manifesting in the idea of the waveguide as a national export product and the Thatcher government's reorganization of System X procurement. But overall, across its projects and places, from TXE4 to London TeleCity, the Post Office prioritized its corporate business.

Rather than welfarist or nationalist, it is thus more accurate to speak of corporate digitalization. The Post Office's corporatization removed it from the Civil Service and replaced the postmaster general, a government minister, with a corporate board. This act distanced the Post Office from political oversight, which was its intended goal. After all, the Post Office's telecommunications business had struggled with government intervention for the best part of a century. But corporatization did little to prevent this intervention, such as Treasury spending controls, which often happened during the 1970s, nor did it protect the Post Office from the IMF's fiscal restraints after the 1976 bailout. Instead, corporatization merely concentrated power among the technocrats that ran the Post Office. This concentration was not a "bug" but a feature. State-owned corporations have always been quasi-independent cadres of professional bureaucrats that manage and deliver specific goods or services.[10] It is thus always a risk that these bureaucrats run their corporations in ways that do not necessarily align with the public or national interest.

This does not mean that the Post Office's corporatization should be seen as necessarily separating it from national or public interest. The Post Office, for example, was committed to its role as a public service in defending the uniformity principle to the City of London up until liberalization, when BT began the London TeleCity project. Some of the Post Office corporation's actions during public ownership came as part of the UK's wider economic nationalism. For example, the Heath and Thatcher governments deferred to the Post Office and BT as the default leaders of Britain's telecom manufacturing industry during the 1970s and 1980s. But the market also motivated the

Post Office throughout this period, even before corporatization. This was not a sharp turn from nationalized public service to privatized corporation. With corporatization, the Wilson government did not intentionally marketize the Post Office, but it nevertheless became more attentive to certain markets, such as the City of London. This accelerated a trend that had already existed since at least TAT-1, when the allure of international communication markets meant the Post Office advocated for liberalization in the United States. This complicates narratives about a breakthrough neoliberal political economy under Margaret Thatcher, in which BT's privatization was a harbinger of change. Since BT's sale, the West has seen a trend of "neoliberal corporatization," in which governments use corporatization to embed market-based operating mechanisms in public services, preparing them for privatization.[11] But the Post Office's corporatization predates this trend. It prefigured neoliberal corporatization and shows that privatization was not a rupture, but something that built on longer and earlier infrastructural change.

This infrastructural change was not just organizational. It was also, via digitalization, forged through technological change. Managers and engineers used computerization to control equipment supply, reorganize labor, and marketize staff. Data networks created new markets, both nationally and internationally, which fiber-optic cables and rooftop satellites helped capture. All these projects helped BT maintain and extend its commercial reach across privatization, by cutting labor and equipment costs and opening new markets. Martlesham Heath, the North Atlantic, and the City of London reveal how much internationalization, in particular, mattered to this changing political economy. These places abetted the internationalization of British telecommunications, shifting focus not just from the government to the corporation but from the national sphere to the international. This doesn't mean that everything went the Post Office and BT's way. The key feature of the digital vision, an integrated digital network, never became a reality. But its failure also shows the limits of economic nationalism. While government was happy to support the Post Office's monopsony over equipment suppliers in the 1970s, for example, it didn't help the Post Office expand its monopoly to television transmission. Instead, digitalization helped the Conservative government in 1991 open telecommunications to competition from cable TV operators. Both in the ways that technology thus supported and undermined the Post Office and BT's monopolies and monopsonies, digitalization, like corporatization, built privatization from the bottom up.

This book does not condemn corporatization or digitalization. Corporatization and digitalization did not determine this history, and both have alternative histories and political economies. Japan, Chile, and France all built alternative political economies of digitalization.[12] Progressive corporatization, rather than neoliberal corporatization, is possible.[13] For example, Électricité de France was founded in 1946 as a public corporation with a strong progressive, left-wing agenda and was dominated by a communist labor union.[14] But this book shows how, in this case, corporatization, digitalization, and privatization combined over three decades to shape a unique history and direction for Britain's telecom infrastructure. It is hard to say how things would have been different if the Post Office, for example, had remained a Civil Service department, closer to democratic, ministerial oversight. Given the power of Britain's government machine among Post Office engineers, remaining a Civil Service department may have only empowered engineers' bureaucratic approach to public digital infrastructure. What is more important to recognize, however, is that successive British governments never gave enough consideration to alternative forms of ownership, whether public, private, or a hybrid of the two.[15] The shift between public and private corporation is but one history of the full range of available alternatives, which is an essential direction for future research. One valuable, untold alternative history is that of Kingston-Upon-Hull's telecom network, the only municipally owned telecom network in the UK. Founded in 1902 as part of the Hull Corporation during a national vogue for municipal socialism, the network remained in public ownership until 1999, and Hull remains the only place in the UK not served by BT.

There is also much work still to be done on a global technological history of privatization. BT's privatization may have been the start, but it was by no means the end. In the UK, British Gas's privatization was also central to "popular capitalism," while privatizing the water supply remains one of the most radical privatizations in history. In Europe, an infrastructural history of privatization can help understand the seeming paradox of how European integration accelerated when the infrastructures that formed the material foundation for that integration were increasingly exiting state ownership. Eastern Europe is a crucial case here, as it became a key destination for the "travelling technocrats" who sold their experience of Western privatizations as advice to new governments after the collapse of the USSR.[16] New Zealand is also a useful case as a country that, like the UK, privatized hard and fast at first but, unlike the UK, managed to reverse course, renationalizing rail, for example. Denationalization has also been a key economic trend in the global south, where the IMF and the World Bank have played a key role in pressuring for privatization, which they see as a way of not just accelerating economic growth but also mitigating against supposed "state failure."[17]

This book shows, however, that infrastructure gives the state a hidden depth that goes beyond changes in ownership. Understanding how managers, engineers, and technology influenced privatization can help undermine the top-down narratives about privatization told by politicians, think tanks, and international organizations. Whether promulgated by Margaret Thatcher's Conservative governments, Jacques Delors' European Commission, or any other national or supranational government or agency, privatization and liberalization are not sharp transformations. To be sure, they are state transformations, rather than simple "hollowing out," but they are longer and slower transformations that emerge from infrastructural, rather than regulatory, change. Looking to infrastructure as the state's hidden depths shows how states take new shapes, in both national and international arenas, before and after ownership and market re-regulation. This means that the histories of these transformations, such as the UK's denationalization, Western Europe's integration, Eastern Europe's post-Soviet transformation, and more, can only be fully understood through the history of the infrastructural change within and across the states participating in those transformations.

These histories would be more than histories of the specific materials and technologies that compose these infrastructures, such as water or rail. They would also all be histories of digitalization too. Digitalization acts as a "universal solvent," and in the period that privatization has dominated global political economy, digitalization gave new ways to model, manage, and control the people and material that composed these infrastructures.[18] Digitalization has also offered key tools that have given new ways for the state to regulate privatized, liberalized infrastructures, from cellular radio spectrum auctions to electricity grid balancing. A global, technological history of privatization and liberalization is thus necessarily also a history of digitalization and its influence on these global transformations.

This book might seem pessimistic in some ways. It shows that there are no shortcuts to transforming infrastructure. Top-down changes in ownership are neither quick nor effective. Instead, the builders and maintainers of infrastructure contest, adapt to, and mutate changes in ownership from the bottom up in ways that those at the top cannot anticipate. The technological momentum imparted by these builders and maintainers is hard to steer, so any ownership change cannot ignore that momentum but instead must work with it.[19] This is particularly true in mature infrastructure, where momentum accumulates over decades or even centuries. Mature infrastructure is thus a conservative force, not necessarily politically, but in the sense that it conserves this technological momentum and the politics that suffuse it. It is easier to load new infrastructure with politics, just as engineers loaded the nascent internet with an ideology of openness or Chilean technologists loaded Project Cybersyn with democratic socialist values. And yet, the solution cannot simply be to build new infrastructures to replace the old. That is unsustainable.

If there is hope for the future of infrastructure, then the history of Britain's telecom infrastructure shows that hope in two ways. First, while mature infrastructure can conserve old politics, it can also inspire new politics. Britain's telecom infrastructure inspired the "London ideology," born before privatization from the decade-long negotiations between the Post Office and the City of London. Second, engineers were responsive to their economic and intellectual environment. By the time of BT's sale, many senior managers and engineers had already wanted privatization for several years and developed technologies and plans that supported privatization. They were convinced not just by a decade of frustrating corporate ownership and financial restrictions, but also by the new ideas about digitalization and information that appeared from the 1950s to the 1970s, promising an alchemical transformation of society. Mature infrastructures respond to new ideas and technologies and can mutate them into new politics. So long as we recover forgotten ideas and inspire new ones, hope remains to change infrastructure for the better.

NOTES

INTRODUCTION

1. Parker, *Official History of Privatisation*, 1: The Formative Years, 1970–1987: 291–316.

2. Veljanovski, *Selling the State*; Bishop and Thompson, "Privatisation in the UK"; Ramanadham, *Privatisation: A Global Perspective*; Stevens, "Evolution of Privatisation,"; Parker, *Official History of Privatisation*; Prasad, *Politics of Free Markets*.

3. Margaret Thatcher, "The Principles of Thatcherism" (Taiwan, September 1, 1992), 108301, Thatcher MSS (Digital Collection), http://www.margaretthatcher .org/document/108301; Margaret Thatcher, "The Principles of Thatcherism" (Seoul, September 3, 1992), 108302, Thatcher MSS (Digital Collection), http://www.marga retthatcher.org/document/108302.

4. Baker, *Turbulent Years*, 84; Margaret Thatcher, "The Principles of Thatcherism" (Seoul, September 3, 1992), 108302, Thatcher MSS (Digital Collection), http://www .margaretthatcher.org/document/108302.

5. "'Crazed Communist Scheme,' PM Johnson Says of Corbyn's Plan for BT"; McRae, "Labour Plans Broadband Communism!"

6. "Telecom Policy" (July 19, 1982), HC Deb Vol. 28 cc23–32, Hansard.

7. Margaret Thatcher, "Speech Opening Conference on Information Technology" (The Barbican Centre, London, December 8, 1982), 105067, Thatcher MSS (Digital Collection), http://www.margaretthatcher.org/document/105067.

8. Kenneth Baker, "Towards an Information Economy" (September 7, 1982), T 471/45, The National Archives (henceforth TNA).

9. Merriman, "Men, Circuits and Systems."

10. Egan, "A Separate Company for a Broadband Network."

11. Crawford, *Captive Audience*; Goldsmith and Wu, *Who Controls the Internet?*

12. Millward, "European Governments and the Infrastructure Industries"; Millward, *Private and Public Enterprise*; Millward, "Business and Government in Electricity Network Integration"; Millward, "Geo-Politics versus Market Structure."

13. Stevens, "Evolution of Privatisation."

14. Foreman-Peck and Millward, *Public and Private Ownership*, 331–338; Parker, *Official History of Privatisation*, 1: The Formative Years, 1970–1987:40; Stevens, "Evolution of Privatisation"; Clarke, "Political Economy of the UK Privatization Programme"

15. Parker, *Official History of Privatisation*, 1: The Formative Years, 1970–1987:52.

16. Conservative Party, "Conservative General Election Manifesto 1979" (London, 1979), https://www.margaretthatcher.org/document/110858; Ortolano, *Thatcher's Progress*.

17. Parker, *Official History of Privatisation*, 1: The Formative Years, 1970–1987:54, 79–81.

18. "British Telecom: Offer for Sale of Ordinary Shares" (1984), TCE 49/2, BT Archives (henceforth BTA); Parker, *Official History of Privatisation*, 1: The Formative Years, 1970–1987:301

19. Parker, *Official History of Privatisation*, 1: The Formative Years, 1970–1987:291–316.

20. Lawson, *View from No. 11*, 224.

21. Stevens, "Evolution of Privatisation."

22. Prasad, *Politics of Free Markets*, 99–103, 131–135.

23. Vinen, *Thatcher's Britain*; Wincott, "Thatcher."

24. Green, "Thatcherism"; Green, *Ideologies of Conservatism*; Thompson, "Thatcherite Economic Legacy"; Saunders, "'Crisis? What Crisis?'"

25. Megginson and Netter, "From State to Market."

26. Myddelton, "British Approach to Privatisation"; Clifton, Comín, and Fuentes, "Privatizing Public Enterprises."

27. Ledger, "Neo-Liberal Thought and Thatcherism."

28. Veljanovski, *Selling the State*; Bishop and Thompson, "Privatisation in the UK"; Ramanadham, *Privatisation: A Global Perspective*; Stevens, "Evolution of Privatisation"; Parker, *Official History of Privatisation*.

29. For "market turn," see Offer, "Market Turn."

30. Venugopal, "Neoliberalism as Concept"; Boas and Gans-Morse, "Neoliberalism"; Ward and England, "Introduction: Reading Neoliberalism."

31. Mirowski and Plehwe, *Road from Mont Pèlerin*; Stedman-Jones, *Masters of the Universe*; Turner, *Neo-Liberal Ideology*; Skousen, *Vienna & Chicago, Friends or Foes?*; Slobodian, *Globalists*.

32. Offer, "Market Turn"; Peck and Tickell, "Conceptualizing Neoliberalism, Thinking Thatcherism"; Peck and Tickell, "Neoliberalizing Space."

33. Venugopal, "Neoliberalism as Concept"; Peck, *Constructions of Neoliberal Reason*; Sutcliffe-Braithwaite, Davies, and Jackson, "Introduction: A Neoliberal Age?"

34. Prasad, *Politics of Free Markets*, 23.

35. Cahill, *End of Laissez-Faire?*; Rollings, "Organised Business and the Rise of Neoliberalism"; Davies, "Roots of Britain's Financialised Political Economy."

36. Offer, "Market Turn"; Berman, *Primacy of Politics*; Davies, *City of London and Social Democracy*.

37. Edgerton, *Rise and Fall*.

38. Edgerton, "What Came between New Liberalism and Neoliberalism?," 45.

39. Billings and Wilson, "'Breaking New Ground,'" 3.

40. Lipartito, "Regulation Reconsidered"; Clifton, Lanthier, and Schröter, "Regulating and Deregulating the Public Utilities 1830–2010."

41. Temin and Galambos, *Fall of the Bell System*; Bauer, "Changing Roles of the State in Telecommunications"; Noam, *Telecommunications in Europe*; Hulsink, *Privatisation and Liberalisation in European Telecommunications*; Thatcher, *Politics of Telecommunications*.

42. Campbell and Pedersen, *Rise of Neoliberalism*; Fourcade-Gourinchas and Babb, "The Rebirth of the Liberal Creed"; Prasad, *Politics of Free Markets*, 23.

43. Pinch, "Technology and Institutions."

44. Edwards, "Infrastructure and Modernity," 191; Hay, "Whatever Happened to Thatcherism?"; Larner and Laurie, "Travelling Technocrats, Embodied Knowledges."

45. John, *Network Nation*; MacDougall, *People's Network*.

46. Agic and Grove, "Role of the State in Telecommunications Infrastructure Financing Across Europe."

47. Fickers and Griset, *Communicating Europe*.

48. Fickers and Griset, *Communicating Europe*.

49. Tworek, *News from Germany*.

50. Griset, "Innovation and Radio Industry."

51. John and Laborie, "'Circuits of Victory.'"

52. Henrich-Franke and Laborie, "European Union for and by Communication Networks"; Laborie, "Fragile Links, Frozen Identities."

53. Campbell-Kelly et al., *Computer*; Ceruzzi, *History of Modern Computing*.

54. Campbell-Kelly and Garcia-Swartz, *From Mainframes to Smartphones*.

55. Campbell-Kelly and Garcia-Swartz, "Economic Perspectives."

56. Campbell-Kelly and Garcia-Swartz, *From Mainframes to Smartphones*.

57. Haigh, "Computing the American Way"; Aspray and Loughnane, "Foreground the Background."

58. Campbell-Kelly, *From Airline Reservations to Sonic*, 6.

59. Usselman, "Unbundling IBM"; Cortada, *IBM*.

60. Hendry, *Innovating for Failure*; Campbell-Kelly, *ICL*; Mounier-Kuhn, "French Computer Manufacturers"; Mounier-Kuhn, "On the History of the Data Processing Industry."

61. Schlombs, *Productivity Machines*.

62. Sumner, "Defiance to Compliance."

63. Hicks, *Programmed Inequality*.

64. Agar, *Government Machine*.

65. Abbate, *Inventing the Internet*.

66. Mailland and Driscoll, *Minitel*.

67. Peters, *How Not to Network*.

68. Medina, *Cybernetic Revolutionaries*.

69. Garvey, "Artificial Intelligence and Japan's Fifth Generation."

70. Haigh, "Inventing Information Systems"; Kline, "Inventing an Analog Past."

71. Kline, *Cybernetics Moment*.

72. Kline, *Cybernetics Moment*, 234–235; Agar, *Government Machine*; Turner, *From Counterculture to Cyberculture*.

73. Winner, "Mythinformation in the High-Tech Era"; Barbrook and Cameron, "Californian Ideology"; Turner, *From Counterculture to Cyberculture*.

74. Rankin, *People's History of Computing*.

75. Edwards, "Infrastructure and Modernity," 216–218.

76. Russell, *Open Standards and the Digital Age*.

77. Russell and Schafer, "In the Shadow of ARPANET"; Henrich-Franke, "EC Competition Law and the Idea of 'Open Networks' (1950s–1980s)."

78. Dewandre, "Europe and New Communication Technologies"; Thatcher, *Internationalisation and Economic Institutions*.

79. Hills, *Deregulating Telecoms*; Temin and Galambos, *Fall of the Bell System*; Cantor and Cantor, "Regulation and Deregulation."

80. Hills, *Deregulating Telecoms*; Cawson et al., *Hostile Brothers*.

81. Thatcher, *Internationalisation and Economic Institutions*.

82. Campbell-Kelly and Garcia-Swartz, "History of the Internet"; Russell, "Histories of Networking"; Russell, *Open Standards*; Balbi and Berth, "Towards a Telephonic History"; Henrich-Franke, "Computer Networks on Copper Cables."

83. Haigh, "Introducing the Early Digital"; Ensmenger, "Digital Construction of Technology"; McGee, "Stating the Field."

84. Misa, "Understanding 'How Computing Has Changed the World'"; Haigh, "History of Information Technology"; Bory, Negro, and Gabriele, "Introduction"; Hu and Peters, "A Conversation."

85. John, *Network Nation*; MacDougall, *People's Network*.

86. Usselman, "Unbundling IBM."

87. Medina, *Cybernetic Revolutionaries*; Garvey, "Artificial Intelligence and Japan's Fifth Generation."

88. John, *Spreading the News*; Joyce, *State of Freedom*.

89. Hughes, *Networks of Power*.

90. Tynan, "Mill and Senior"; Mosca, "On the Origins of the Concept of Natural Monopoly"; Plaiss, "From Natural Monopoly to Public Utility."

91. Usselman, *Regulating Railroad Innovation*.

92. Kaijser, "Helping Hand."

93. Guldi, *Roads to Power*.

94. Badenoch and Fickers, *Materializing Europe*; Högselius, Kaijser, and van der Vleuten, *Europe's Infrastructure Transition*; Barry, *Political Machines*.

95. Lengwiler, "Technologies of Trust."

96. Seely, *Building the American Highway System*; Light, *From Warfare to Welfare*.

97. Ortolano, *Thatcher's Progress*.

98. Pellizzoni and Ylönen, *Neoliberalism and Technoscience*.

99. Mirowski, *Machine Dreams*; Mirowski and Nik-Khah, *Knowledge We Have Lost in Information*.

100. Mitchell, *Carbon Democracy*.

101. Hecht, *Radiance of France*; Edgerton, *England and the Aeroplane*; Elam, "National Imaginations"; Edgerton, "Contradictions of Techno-Nationalism and Techno-Globalism."

102. Alder, *Engineering the Revolution*; Hecht, *Radiance of France*; Edwards and Hecht, "History and the Technopolitics of Identity"; Karas and Arapostathis, "Harbours of Crisis and Consent."

103. Mitchell, *Rule of Experts*.

104. Barry, "Anti-Political Economy."

105. Ortolano, *Thatcher's Progress*, 24; Edwards, "Infrastructure and Modernity," 220–223.

106. Bowker, "Information Mythology," 235; Bowker and Star, *Sorting Things Out*, 33–50.

107. Hughes, "Electrification of America"; Hughes, *Networks of Power*.

108. Chandler, *Visible Hand*; Galambos, "Technology, Political Economy, and Professionalization."

109. For an example of such conflict, see Akera, "Engineers or Managers?"

110. Yates, *Control through Communication*; Agar, *Government Machine*.

111. Hughes, *Networks of Power*; Hecht, *Radiance of France*.

112. Boltanski and Chiapello, *New Spirit of Capitalism*; Kohlrausch and Trischler, *Building Europe on Expertise*; Mitchell, *Rule of Experts*.

113. Hecht, "Planning a Technological Nation"; Law, "Technology and Heterogeneous Engineering."

114. Vernon, "The Local, the Imperial and the Global."

115. Thatcher, *Internationalisation and Economic Institutions*.

116. Edwards, *Closed World*.

117. Abbate, *Inventing the Internet*, 147–181.

118. Pitt, *Telecommunications Function in the British Post Office*, 25–27; Vernon, "The Local, the Imperial and the Global," 416; Cronin, *Politics of State Expansion*; Bradshaw, "Dead Hand of the Treasury."

119. Hennessy, *Whitehall*, 69, 171, 181–182; Edgerton, *Science, Technology and the British Industrial "Decline", 1870–1970*; Agar, *Government Machine*, 198–199, 308, 411.

120. Exceptions include Bruton and Gooday, "Listening in Combat"; McGuire, "Categorisation of Hearing Loss"; Haigh, "'To Strive, to Seek, to Find'"; Boon, "'Research Is the Door to Tomorrow.'"

121. Campbell-Smith, *Masters of the Post*, 176–179, 193.

122. Campbell-Smith, *Masters of the Post*, 197; Pitt, *Telecommunications Function in the British Post Office*, 37. This takeover excluded the city of Kingston-upon-Hull, which ran its own network as the last remaining holdover from municipal socialist experiments with city-owned networks in the 1880s and 1890s.

123. Campbell-Smith, *Masters of the Post*, 299–301; Pitt, *Telecommunications Function in the British Post Office*, 71, 78–82, 102.

124. Post Office, *Report on Post Office Development and Finance*.

125. Pitt, *Telecommunications Function in the British Post Office*, 141; Post Office, *Inland Telephone Service*.

126. "Freedom for the GPO"; "All Change at the Post Office"; "A Commercial Post Office."

127. "Post Office Report and Accounts" (1971), TCC 11/2, BTA; "Post Office Report and Accounts" (1972), TCC 11/3, BTA.

128. Thatcher, *Politics of Telecommunications*, 123–124.

129. Post Office Review Committee, *Report of the Post Office Review Committee*.

130. Department of Industry, *Post Office*.

131. Campbell-Smith, *Masters of the Post*, 551–552.

132. Hamilton, "Joseph to Split Post Office."

133. Parker, *Official History of Privatisation*, 1: The Formative Years, 1970–1987:243.

134. Beesley, *Liberalisation of the Use of British Telecommunications Network*.

135. Parker, *Official History of Privatisation*, 1: The Formative Years, 1970–1987:241–250.

136. "Telecom Policy" (July 19, 1982), HC Deb Vol 28 cc23–32, Hansard.

137. Kline, *Cybernetics Moment*; Borup et al., "Sociology of Expectations"; Messeri and Vertesi, "Greatest Missions Never Flown."

138. McCray, *Visioneers*; Andersson, "Great Future Debate"; Andersson and Rindzevičiūtė, *Struggle for the Long-Term*; Andersson, *Future of the World*.

139. Vernon, "The Local, the Imperial and the Global"; Ortolano, *Thatcher's Progress*; Brooke, "Living in 'New Times'"; Edwards, "Infrastructure and Modernity"; Larner and Laurie, "Travelling Technocrats, Embodied Knowledges."

CHAPTER 1

1. Merriman, "Men, Circuits and Systems," 247.

2. Merriman, "Men, Circuits and Systems," 241.

3. Merriman, "Men, Circuits and Systems," 250.

4. Gooday, "Re-writing the 'Book of Blots'"; Lipartito, "Picturephone and the Information Age."

5. Borup et al., "Sociology of Expectations"; McCray, *Visioneers*; Fjaestad, "Fast Breeder Reactors in Sweden."

6. Gordon Radley to Ernest Marples, "Electronic Switching," April 12, 1957, TCB 2/113, BTA.

7. Harris, *Automatic Switching in the UK*, 52.

8. Huurdeman, *Worldwide History of Telecommunications*, 193–194; Liffen, "Telegraphy and Telephones."

9. "Report of an Official Visit to the United States of America to Study Developments in Telephone Switching Practice" (1947), TCB 371/39, BTA.

10. Harris, *Automatic Switching in the UK*, 53.

11. Boon, "'Research Is the Door to Tomorrow.'"

12. Randell, "Of Men and Machines."

13. Copeland, *Colossus*; Haigh and Priestley, "Colossus and Programmability," 23.

14. Fensom, "How Colossus Was Built," 301; Copeland, "Machine against Machine," 76.

15. Gordon Radley to Ernest Marples, "Electronic Switching," April 12, 1957, TCB 2/113, BTA.

16. Harris, *Automatic Switching in the UK*, 53.

17. Edgerton, *Warfare State*.

18. Flowers, "D-Day at Bletchley Park," 83.

19. Flowers, "D-Day at Bletchley Park," 83.

20. Copeland, "Colossus and the Rise," 109.

21. L. R. F. Harris, "Highgate Wood: General System Description" (December 1960), HIC W04/01 Switching/Public Automatic Exchanges/Experimental and Prototype Systems, BTA.

22. Gordon Radley to Ernest Marples, "Electronic Switching," April 12, 1957, TCB 2/113, BTA.

23. Gordon Radley to Ernest Marples, "Electronic Switching," April 12, 1957, TCB 2/113, BTA.

24. Gordon Radley to Ernest Marples, "Electronic Switching," April 12, 1957, TCB 2/113, BTA.

25. "Minutes of a Meeting of the Joint Electronic Research Committee" (September 2, 1957), TCB 2/113, BTA.

26. J. Bellew to O. H. Lawn, "Tudor T.E. 2nd Unit (Highgate Wood Exchange)," August 2, 1957, POST 122/894, BTA.

27. Agar, *Science and Spectacle*.

28. Bud, "Penicillin and the New Elizabethans."

29. Harris, *Automatic Switching in the UK*, 10.

30. Harris, *Automatic Switching in the UK*, 10–11.

31. Harris, *Automatic Switching in the UK*, 11.

32. "Consultative Group on Electronic Exchange Developments: The Future of Highgate Wood" (January 1964), TCB 2/113, BTA.

33. "An Appraisement of an Integrated P.C.M. System" (July 1963), TCB 422/20854, BTA.

34. "The Possibility of Progressive Conversion of a Telephone Area from an Analogue to an Integrated P.C.M. System, Part 1" (April 1964), TCB 422/20953, BTA.

35. "Rationalisation of the Distribution Network: First Meeting of the Working Party" (November 16, 1965), TCB 807/751, BTA; "Report of the Working Party on the Rationalisation of the Distribution Network" (January 31, 1966), TCB 807/751, BTA.

36. "Post Office Board Meeting" (April 24, 1967), TCB 14/5, BTA; "Rationalisation of the Local Distribution Network" (October 1968), TCB 54/2/45, BTA.

37. "The Possibility of Progressive Conversion of a Telephone Area from an Analogue to an Integrated P.C.M. System, Part 2" (August 1967), TCB 422/20953, BTA.

38. "Post Office Board Meeting" (April 24, 1967), TCB 14/5, BTA; "Post Office Board Meeting" (July 5, 1967), TCB 14/7, BTA.

39. "Post Office Board Meeting" (July 22, 1968), TCB 14/14, BTA; "Rationalisation of the Local Distribution Network" (October 1968), TCB 54/2/45, BTA.

40. Huurdeman, *Worldwide History of Telecommunications*, 327–331.

41. "The World's First PCM Exchange"; Kerswell and Jones, "Conclusions from the Empress Digital Tandem Exchange."

42. Wiener, *Cybernetics*; Shannon, "Mathematical Theory of Communication, Part 1"; Shannon, "Mathematical Theory of Communication, Part 2."

43. Kline, *Cybernetics Moment*.

44. Kline, *Cybernetics Moment*, 68–102, 135–151.

45. Kline, *Cybernetics Moment*, 102–134.

46. "An Introduction to Information Theory" (August 1951), TCB 422/13450, BTA; Kline, *Cybernetics Moment*, 104.

47. "A General Introduction to Communication Theory" (1951), IPOEE Unpublished Papers 1950–52, Part 3, HIC 001/001/0015, BTA.

48. "An Introduction to Information Theory" (August 1951), TCB 422/13450, BTA.

49. "An Introduction to Information Theory" (August 1951), TCB 422/13450, BTA.

50. Kline, *Cybernetics Moment*, 52–55.

51. "An Introduction to Information Theory" (August 1951), TCB 422/13450, BTA.

52. Kline, *Cybernetics Moment*, 117.

53. "Notes on a Symposium on Communication Theory Arranged by Imperial College" (December 1952), TCB 226/2227, BTA.

54. "Notes on a Symposium on Communication Theory Arranged by Imperial College" (December 1952), TCB 226/2227, BTA.

55. "An Introduction to Information Theory" (August 1951), TCB 422/13450, BTA.

56. "An Introduction to Information Theory" (August 1951), TCB 422/13450, BTA; "A General Introduction to Communication Theory" (1951), IPOEE Unpublished Papers 1950–52, Part 3, HIC 001/001/0015, BTA.

57. Merriman, "Men, Circuits and Systems," 241.

58. Merriman, "Men, Circuits and Systems," 241.

59. Merriman, "Men, Circuits and Systems," 242.

60. "Appointment of Mr. J.H.H. Merriman as Senior Director of Engineering."

61. Agar, *Government Machine*.

62. Agar, *Government Machine*, 424–430.

63. Agar, *Government Machine*, 330.

64. For more on O&M-GPO ties, see Agar, *Government Machine*, 427.

65. Agar, *Government Machine*, 258.

66. Merriman and Wass, "To What Extent Can Administration Be Mechanized?"

67. Agar, *Government Machine*, 69–74.

68. Agar, *Government Machine*, 397.

69. Merriman, "Men, Circuits and Systems," 248.

70. Edwards, *Closed World*; Head, "Getting Sabre Off the Ground."

71. Yates, *Turing's Legacy*, 128–129; Abbate, *Inventing the Internet*, 7–42.

72. Campbell-Kelly, *From Airline Reservations to Sonic the Hedgehog*, 63; Yost, *Making IT Work*, 166.

73. Abbate, *Inventing the Internet*, 34–35.

74. Abbate, *Inventing the Internet*, 29.

75. "Interactions Between Data Processing and Telecommunications" (June 1971), TCC 55/3/76, BTA.

76. Abbate, *Inventing the Internet*, 29, 35; Gillies and Cailliau, *How the Web Was Born*, 55–56; Tuomi, *Networks of Innovation*, 75; Agar, *Government Machine*, 381.

77. Baran, "Future Computer Utility."

78. Pitt, *Telecommunications Function in the British Post Office*, 141; Campbell-Smith, *Masters of the Post*, 423.

79. Post Office, *Inland Telephone Service*.

80. Pitt, *Telecommunications Function in the British Post Office*, 145.

81. Caves, *Britain's Economic Prospects*; Pitt, *Telecommunications Function in the British Post Office*, 146; "Freedom for the GPO"; "All Change at the Post Office"; "A Commercial Post Office."

82. Whyte, *Organization Man*; McKenna, *World's Newest Profession*.

83. Pitt, *Telecommunications Function in the British Post Office*, 137–138.

84. Agar, *Government Machine*, 330–331; Lowe, "Milestone or Millstone?"

85. Pitt, *Telecommunications Function in the British Post Office*, 140.

86. "Introducing You to LEAPS."

87. "Post Office Enters the Computer Age"; "Computer Centre in Kensington."

88. "Progress: Kensington Computer Centre" (1965), TCB 420/IRP (PR) 4, BTA.

89. Edgerton, "'White Heat' Revisited."

90. For more on Benn's tenure as postmaster general, see Campbell-Smith, *Masters of the Post*, 438–469.

91. "Annex A: Copy of a Minute from the Postmaster General to the Prime Minister. Organisation of the Post Office" (March 2, 1965), PREM 13/1063, TNA.

92. "Mr Hunt: P.O. (W.P.)(65)4" (June 11, 1965), T 319/167, TNA.

93. "Organisation of the Post Office: Report by a Working Party of Officials" (July 1965), PREM 13/1063, TNA.

94. "Prime Minister: Post Office Organisation" (July 20, 1965), PREM 13/1063, TNA.

95. "Post Office Re-Organisation: Memorandum by the Postmaster General" (August 1965), PREM 13/1063, TNA.

96. "Steering Group on the Organisation of the Post Office: Report of the Steering Group" (January 18, 1966), T 319/173, TNA.

97. "ED(PO)66 2nd Meeting. Cabinet Ministerial Committee on Economic Development. Sub-Committee on Status and Organisation of the Post Office" (February 10, 1966), CAB 134/1761, TNA.

98. Campbell-Smith, *Masters of the Post*, 470.

99. "House of Commons. Minutes of Evidence Taken before the Select Committee on Nationalised Industries. Meeting at Post Office Research Station, Dollis Hill" (July 12, 1966), POST 122/10345, British Postal Museum and Archive (henceforth BPMA).

100. *First Report from the Select Committee on Nationalised Industries: The Post Office*, Volume 1: Report and Proceedings of the Committee:205

101. *First Report from the Select Committee on Nationalised Industries: The Post Office*, Volume 1: Report and Proceedings of the Committee:205.

102. Campbell-Smith, *Masters of the Post*, 461–464.

103. McKenna, *World's Newest Profession*, 166; Roeber, "In This Concluding Article."

104. "McKinsey and Company: Progress Review with Postmaster General" (February 23, 1967), POST 72/906, BPMA.

105. Post Office, *Reorganisation of the Post Office*.

CHAPTER 2

1. I also discuss this department in Ward, "Computer Models and Thatcherist Futures"; Ward, "Nineteen Eighty-Four in the British Telephone System."

2. Harris, *Automatic Switching in the UK*, 16.

3. "Long Range Studies Report 19: City in the Year 2000" (1969), TCC 252/19, BTA.

4. Probert, "Development of a Long-Range Planning Model," 708; Probert, "Systems Dynamics Modelling," 73–74.

5. "Into the 21st Century: Proceedings of a Long Range Strategy Seminar" (1980), 24, TCC 75/1, BTA.

6. "Into the 21st Century: Proceedings of a Long Range Strategy Seminar" (1980), 24, TCC 75/1, BTA.

7. O'Hara, *From Dreams to Disillusionment*.

8. Andersson, "Great Future Debate"; Andersson, *Future of the World*.

9. Andersson, *Future of the World*, 35–36.

10. Andersson, *Future of the World*, 50.

11. Andersson, *Future of the World*, 220.

12. Messeri and Vertesi, "Greatest Missions Never Flown."

13. Harris, *Automatic Switching in the UK*, 16.

14. "McKinsey and Company: Progress Review with Postmaster General" (February 23, 1967), POST 72/906, BPMA.

15. "House of Commons. Minutes of Evidence Taken before the Select Committee on Nationalised Industries. Meeting at Post Office Research Station, Dollis Hill" (July 12, 1966), 7–8, POST 122/10345, BPMA.

16. "Telecommunications System of the Future" (1967), TCB 662/1, BTA; "Telecommunications Systems of the 1980s" (1967), TCB 662/2, BTA.

17. Harris, *Automatic Switching in the UK*, 16.

18. Gabor, *Inventing the Future*, 207.

19. "Notes on a Symposium on Communication Theory Arranged by Imperial College" (December 1952), TCB 226/2227, BTA; "Computer Simulation of Professor Gabor's Speech Compression System" (December 1966), TCB 422/21230, BTA.

20. "A Presentation of the Work of the United Kingdom Trunk Task Force" (1969), TCC 145/1, BTA; "AGSD Report No. 1: Costs Arising from Variations in the Design of Telephone Exchange Equipment" (September 1970), TCC 55/3/45, BTA.

21. For more on ICL, see Campbell-Kelly, *ICL*.

22. *Telecommunication Services for the 1990s*; "Telecommunications Services for the 1990s"; "Telecommunications System of the Future" (1967), TCB 662/1, BTA;

"Telecommunications Systems of the 1980s" (1967), TCB 662/2, BTA; "A Presentation of the Work of the United Kingdom Trunk Task Force" (1969), TCC 145/1, BTA; "UK Trunk Task Force Final Report, Volume 1: Guide and Summary Chapters" (1971), TCC 145/7, BTA.

23. "Long Range Studies Report 8: A Marketing and Technical Appreciation of Viewphone" (1969), TCC 252/8, BTA.

24. "The Viewphone Trial—Questionnaire Results" (1976), TCC 23/563, BTA.

25. "Bill's a Winner"; Nuttall, "Well, Hello."

26. Lipartito, "Picturephone and the Information Age," 77.

27. "A Presentation of the Work of the United Kingdom Trunk Task Force" (1969), TCC 145/1, BTA.

28. Breary, "A Long-Term Study of the United Kingdom Trunk Network, Part 1," 214; Breary, "A Long-Term Study of the United Kingdom Trunk Network, Part 2," 2.

29. "UK Trunk Task Force Final Report, Volume 1: Guide and Summary Chapters" (1971), TCC 145/7, BTA; Breary, "A Long-Term Study of the United Kingdom Trunk Network, Part 1"; "Managing Director's Committee: Telecommunications" (November 17, 1972), TCC 55/4/194, BTA.

30. Breary, "A Long-Term Study of the United Kingdom Trunk Network, Part 1," 212.

31. Breary, "A Long-Term Study of the United Kingdom Trunk Network, Part 1," 212.

32. "UK Trunk Task Force Final Report, Volume 1: Guide and Summary Chapters" (1971), TCC 145/7, BTA.

33. "Long Range Studies Report 16: A Review of Premature Obsolescence and Depreciation Policy—Part 1" (1970), TCC 252/16, BTA; "Long Range Studies Report 17: A Review of Premature Obsolescence and Depreciation Policy—Part 2" (1970), TCC 252/17, BTA; "Long Range Studies Report 18: A Review of Premature Obsolescence and Depreciation Policy—Part 3" (1970), TCC 252/18, BTA; "Long Range Studies Report 25: A Review of Premature Obsolescence and Depreciation Policy—Part 4" (1970), TCC 252/25, BTA.

34. "Economic Appraisal of Exchange Equipment Strategies by Computer Model" (1972), FV 87/2, TNA; "PO Model for Evaluating Alternative Equipment Strategies" (November 7, 1972), FV 87/6, TNA; "Long Range Studies Report 18: A Review of Premature Obsolescence and Depreciation Policy—Part 3" (1970), TCC 252/18, BTA.

35. "Managing Director's Committee: Telecommunications" (January 8, 1971), TCC 55/3/11, BTA; "Post Office Board Meeting" (April 26, 1971), TCC 15/4, BTA; Clark et al., *Process of Technological Change*, 46.

36. "Long Range Studies Report 19: City in the Year 2000" (1969), TCC 252/19, BTA; "A Panorama of Telecommunications in the Year 2000" (1970), TCC 274/2, BTA; "Changing Characteristics of Telecommunications and Their Influence upon

Society" (1970), TCC 274/3, BTA; Whyte, "Telecommunications"; "Telecommunications in the Service of Man" (1971), TCC 274/4, BTA.

37. Thring and Laithwaite, *How to Invent*.

38. "Long Range Studies Report 19: City in the Year 2000" (1969), TCC 252/19, BTA.

39. Westin, *Privacy and Freedom*; Committee on Privacy, *Report of the Committee on Privacy*.

40. Agar, *Government Machine*, 359–361.

41. "New Chief."

42. New Scientist. "Long Range Intelligence Division, Post Office Telecommunications HQ: Information Scientist."

43. "Long Range Studies Reports" (1968–1973), TCC 252, BTA; "Long Range Studies Memoranda" (1969–1970), TCC 273, BTA; "Long Range Studies Division Reference Papers" (1969–1975), TCC 274, BTA; "Long Range Studies by Research Contracts" (1971–1973), TCC 272, BTA.

44. "LRSR 10, 12, 13: Long Term Profile of the UK, Parts 1–3" (1969), TCC 252/10, 12, 13, BTA; "Long Range Studies Report 19: City in the Year 2000" (1969), TCC 252/19, BTA; "Long Range Studies Report 1000: Britain 2001 AD" (1971), TCC 272/1000, BTA.

45. "Long Range Intelligence Bulletins" (1974–1978), TCC 90, BTA; "Long Range Research Reports" (1975–1977), TCC 92, BTA.

46. "Long Range Intelligence Bulletin 1: Long Range Economic Forecasts: The Economic Consequences of Energy Scarcity" (1974), TCC 90/1, BTA.

47. "Post Office Report and Accounts" (1974), TCC 11/5, BTA; "Post Office Report and Accounts" (1979), TCC 11/10, BTA; "Long Term Economic and Technological Trends" (1974), TCC 55/6/33, BTA.

48. Meadows et al., *Limits to Growth*; for an overview, see Vieille Blanchard, "Modelling the Future"; and Seefried, "Towards the Limits to Growth?"

49. Forrester, "Industrial Dynamics"; Thomas and Williams, "The Epistemologies of Non-Forecasting Simulations, Part I."

50. "Long Range Intelligence Bulletin 1: Long Range Economic Forecasts: The Economic Consequences of Energy Scarcity" (1974), 6, TCC 90/1, BTA.

51. "Long Range Intelligence Bulletin 2: Long Range Social Forecasts: Working from Home" (1974), TCC 90/2, BTA; "Long Range Intelligence Bulletin 8: Long Range Social Forecasts: Attitudes to Work" (1975), TCC 90/8, BTA.

52. Kahn and Wiener, *The Year 2000*.

53. Bell, *Coming of Post-Industrial Society*.

54. Webster, *Theories of the Information Society*, 38.

55. Andersson, *Future of the World*, 54, 100–103.

56. Probert, "Systems Dynamics Modelling," 69.

57. David Probert, e-mail message to Jacob Ward, "British Telecom Computer Modelling—70s/80s—BT Privatisation," August 30, 2020.

58. Probert, "Systems Dynamics Modelling"; Probert, "Development of a Long-Range Planning Model"; "Strategic Modelling in British Telecom" (1982), TCD 278/PR 42, BTA.

59. Probert, "Development of a Long-Range Planning Model."

60. Probert, "Development of a Long-Range Planning Model."

61. Post Office Review Committee, *Report of the Post Office Review Committee.*

62. Campbell-Smith, *Masters of the Post,* 539.

63. Department of Industry, *Post Office.*

64. "Industrial Democracy 'Test Case'"; "Post Office Democratic Pioneers"; Ardill, "Worker-Directors Dropped by Post Office Management."

65. Williamson, "Bullock Report on Industrial Democracy."

66. Eric Batstone, Anthony Ferner, and Michael Terry, "Interim Report on the Post Office Industrial Experiment: National Level" (1979), POST 65/178, BPMA.

67. Campbell-Smith, *Masters of the Post,* 551; "Post Office Board Meeting" (September 4, 1979), TCC 15/24, BTA.

68. "HC Deb: Post Office Board (Industrial Democracy)" (December 12, 1979), Vol 975 cc1303–1312, Hansard; Campbell-Smith, *Masters of the Post,* 551–552.

69. Hamilton, "Joseph to Split Post Office."

70. Allenstein and Probert, "Strategic Control Module."

71. Probert, "Systems Dynamics Modelling."

72. Foster, "Michael Beesley."

73. Beesley and Littlechild, "Privatisation: Principles, Problems and Priorities"; Beesley and Littlechild, "Regulation of Privatized Monopolies"; Beesley, *Privatization, Regulation and Deregulation.*

74. Beesley, *Liberalisation of the Use of British Telecommunications Network.*

75. Thatcher, *Politics of Telecommunications,* 176.

76. "Future Structure of the Post Office" (March 1979), TCC 55/11/14, BTA; "Telecommunications Board Structure and Organisation" (May 1981), TCC 62/2/42, BTA.

77. "Business Planning and Strategy Department: Report" (1982), TCD 93, BTA.

78. "Strategic Modelling in British Telecom" (1982), TCD 278/PR 42, BTA; Doubleday and Probert, "Development of an Integrated Communications Demand Model."

79. Probert, "Development of a Long-Range Planning Model," 697. Probert was quoting the US Democratic senator for South Carolina, Ernest Hollings, best known for his campaigns against poverty, but Probert's familiarity with Hollings is unclear.

80. Doubleday and Probert, "Development of an Integrated Communications Demand Model," 1092.

81. Doubleday and Probert, "Development of an Integrated Communications Demand Model," 1087.

82. Doubleday and Probert, "Development of an Integrated Communications Demand Model," 1092.

83. Norman Tebbitt to Margaret Thatcher, "POEU Action," October 17, 1983, PREM 19/1344, TNA.

84. "Notes of a Meeting with POEU to Discuss Liberalisation" (April 7, 1983), TCD 69/3/61, BTA.

85. "Telecommunications Board: Arrangements for Liberalisation" (November 25, 1980), TCC 59/1, BTA.

86. Newman, *Selling of British Telecom*, 153.

87. Michael Bett to Board Members, "Privatisation," October 13, 1982, TCD 69/2/115, BTA; "BT and Privatisation: A Situation Report for All Managers and Their Staff" (October 1982), TCD 69/2/115, BTA.

88. Michael Bett, "British Telecom: Assurances for Staff about Privatisation" (April 25, 1983), TCD 69/3/67, BTA; Michael Bett, "Assurances for Staff about Privatisation" (April 25, 1983), TCD 69/3/118, BTA; "Mercury Interconnection and Privatization" (June 21, 1983), TCD 69/3/96, BTA.

89. Mirowski, *Machine Dreams*, 532.

90. Probert, "Development of a Long-Range Planning Model," 703–705; Doubleday and Probert, "Development of an Integrated Communications Demand Model," 1087–1093.

91. Probert, "Development of a Long-Range Planning Model," 708; Probert, "Systems Dynamics Modelling," 73–74.

92. "Business Planning and Strategy Department: Report" (1982), TCD 93, BTA; "The Demand for Residential Information Services" (1981), TCC 75/2, BTA.

93. "Into the 21st Century: Proceedings of a Long Range Strategy Seminar" (1980), TCC 75/1, BTA.

94. "Into the 21st Century," 1.

95. "Into the 21st Century," 77.

96. "Into the 21st Century," 108.

97. "Into the 21st Century," 60.

98. "Into the 21st Century," 4.

99. "Into the 21st Century," 58.

100. "Into the 21st Century," 24.

101. Schumacher, *Small Is Beautiful.*

102. Turner, *From Counterculture to Cyberculture*; Barbrook and Cameron, "Californian Ideology."

103. "Into the 21st Century: Proceedings of a Long Range Strategy Seminar" (1980), 60, TCC 75/1, BTA.

104. "Into the 21st Century," 7.

105. Kenneth Baker, "Towards an Information Economy" (September 7, 1982), T 471/45, TNA.

106. "Telecommunications Bill: 2nd Reading" (November 29, 1982), PREM 19/1100, TNA; "Screen Test."

107. Turner, *From Counterculture to Cyberculture*; Barbrook and Cameron, "Californian Ideology."

108. "Into the 21st Century: Proceedings of a Long Range Strategy Seminar" (1980), 8, TCC 75/1, BTA.

109. "Into the 21st Century," 19.

110. "Into the 21st Century," 103.

111. "Into the 21st Century," 4, 103.

112. "Into the 21st Century," 8.

113. "Into the 21st Century," 23.

114. Kumar, *Utopia and Anti-Utopia*, 221–222.

115. "Into the 21st Century: Proceedings of a Long Range Strategy Seminar" (1980), 23, TCC 75/1, BTA.

116. Lyon, *Electronic Eye*, 57–80; Kumar, *Utopia and Anti-Utopia*, 296.

117. Andersson, *Future of the World*, 224.

118. Bogard, *Simulation of Surveillance*, 4–5.

119. "An Introduction to Information Theory" (August 1951), TCB 422/13450, BTA.

120. Kline, *Cybernetics Moment*, 218, 240.

CHAPTER 3

1. Cawson et al., *Hostile Brothers*, 85.

2. MacDougall, *People's Network*, 153–159.

3. Liffen, "Epsom, Britain's First Public Automatic Telephone Exchange."

4. Cawson et al., *Hostile Brothers*, 77.

5. Noam, *Telecommunications in Europe*, 4.

6. Coopey, "Industrial Policy."

7. Scranton, "None-Too-Porous Boundaries"; Noble, *Forces of Production*.

8. Blok and Downey, *Uncovering Labour*.

9. Downey, "Virtual Webs, Physical Technologies."

10. Downey, *Telegraph Messenger Boys*; Hindmarch-Watson, *Serving a Wired World*; Feigenbaum and Gross, "Answering the Call of Automation"; Feigenbaum and Gross, "Organizational Frictions and Increasing Returns."

11. Russell and Vinsel, "After Innovation, Turn to Maintenance"; Vinsel and Russell, *Innovation Delusion*; Mossop, "Infrastructures of Repair."

12. Bealey, *Post Office Engineering Union*.

13. Light, "When Computers Were Women"; Margot Lee Shetterly, *Hidden Figures*; Haigh, "Inventing Information Systems"; Ensmenger, *Computer Boys Take Over*; Misa, *Gender Codes*; Abbate, *Recoding Gender*; Ensmenger, "'Beards, Sandals, and Other Signs'"; Hicks, *Programmed Inequality*.

14. Schafer and Thierry, *Connecting Women*.

15. Fischer, "Touch Someone"; Fischer, *America Calling*; Dalibert and De Iulio, "Representational Intertwinement."

16. Martin, *"Hello Central?"*; Green, "Goodbye Central"; Lipartito, "When Women Were Switches."

17. Green, "Race, Gender, and National Identity"; Green, *Race on the Line*.

18. *"GRACE" in Action*; *Queen Dials Edinburgh*; Barron, "Dialling for Trunk Calls"; Post Office, *Full Automation of the Telephone System*; Post Office, *Telephone Policy: The Next Steps*.

19. Pitt, *Telecommunications Function in the British Post Office*, 102–104.

20. Pitt, *Telecommunications Function in the British Post Office*, 152.

21. "Prime Minister: Post Office Finance (C.P. (55) 146)" (October 17, 1955), PREM 11/2459, TNA; Post Office, *Report on Post Office Development and Finance*.

22. Kline, *Cybernetics Moment*, 70.

23. "Age of Automation"; Wilson, "Labour's Plan for Science."

24. Booth, *Management of Technical Change*, 8; Edgerton, *Shock of the Old*, 85.

25. "Girl with the Golden Voice."

26. Agar, *Government Machine*, 427.

27. Agar, *Government Machine*, 427–429.

28. Smith, "So These Are 'Bits.'"

29. Post Office, *Telephone Policy: The Next Steps.*

30. Post Office, *Telephone Service and the Customer,* ii.

31. Post Office, *Telephone Service and the Customer;* "HC Deb: Telephone Service" (March 11, 1959), Vol 601 cc1260–1263, Hansard.

32. "Union Approves the 'Friendly Telephone.'"

33. Green, "Race, Gender, and National Identity."

34. Harris, *Automatic Switching in the UK,* 18.

35. W. J. Bray to J. H. H. Merriman, "Project 'ADMITS,'" November 3, 1967, TCB 712/27/2, BTA.

36. "Proposals for a Research and Development Program Leading to an Evolutionary Telecommunications System: Project 'ADMITS'" (November 1967), TCB 712/27/2, BTA.

37. James Merriman to William Bray, "ADMITS", November 20, 1967, TCB 712/27/2, BTA.

38. "A Presentation of the Work of the United Kingdom Trunk Task Force" (1969), TCC 145/1, BTA; "AGSD Report No. 1: Costs Arising from Variations in the Design of Telephone Exchange Equipment" (September 1970), TCC 55/3/45, BTA.

39. Harris, *Automatic Switching in the UK,* 10–11.

40. Sheppard, "The Leighton Buzzard Electronic Telephone Exchange."

41. Cortada, *Digital Hand,* Volume 1: How Computers Changed the Work of American Manufacturing, Transportation, and Retail Industries:169; Campbell-Kelly et al., *Computer,* 48; Clarke and Tobias, "Complexity in Corporate Modelling."

42. Bowker, "How to Be Universal"; Thomas and Williams, "Epistemologies of Non-Forecasting Simulations, Part I"; Vieille Blanchard, "Modelling the Future"; Edwards, *A Vast Machine;* Ward, "Computer Models and Thatcherist Futures."

43. Thomas, *Rational Action;* Light, *From Warfare to Welfare;* Turnbull, "Simulating the Global Environment."

44. Hecht, *Radiance of France,* 102–111.

45. Agar, *Government Machine.*

46. "Telephone Exchange Switching: Modernization Strategy" (1971), TCC 387/ THQ ICU 45(a), BTA.

47. "Post Office Switching Policy: Meeting between the Post Office, GEC and Plessey" (January 12, 1972), FV 87/1, TNA.

48. Cole, "Computer Exchange."

49. "Post Office Switching Policy: Meeting between the Post Office, GEC and Plessey" (February 23, 1972), FV 87/2, TNA.

50. "Post Office Switching Policy: Meeting between the Post Office, GEC and Plessey" (January 12, 1972), FV 87/1, TNA; "Economic Appraisal of Exchange Equipment Strategies by Computer Model" (1972), FV 87/2, TNA.

51. "MPT Report on Discussions between the Post Office and GEC/Plessey" (May 16, 1972), FV 87/1, TNA.

52. "Ryland's Row"; "Fight for the New Telephone System"; "Post Office at Grips with Suppliers"; "Post Office Equipment Row."

53. "Meeting at Post Office Central Headquarters" (September 22, 1972), FV 87/2, TNA.

54. "Post Office Telephone Exchange Equipment: Meeting in the Treasury with Plessey" (November 29, 1972), FV 87/2, TNA.

55. Ken Corfield to Frank Wood, "Questions about TXE4," February 17, 1972, FV 87/1, TNA.

56. "Meeting between the Prime Minister and Senior Manufacturing Industry Managers" (May 25, 1972), FV 87/2, TNA.

57. John Eden to Edward Heath, "Report on Morton's Complaint," June 14, 1972, FV 87/2, TNA.

58. Agar, *Government Machine*, 424.

59. "Long-Term Investment in Exchange Equipment: A Report to Plessey on the Post Office Investment by T.S. Barker, Senior Research Officer, Department of Applied Economics, University of Cambridge" (December 1972), FV 87/6, TNA.

60. "Commentary by P.O. on Report to Plessey by T. S. Barker, Department of Applied Economics, University of Cambridge" (February 6, 1973), FV 87/6, TNA.

61. It is unclear whether this is the Stephen Littlechild who would go on to become an influential advocate for utility privatization and deregulation during the 1980s, but it seems likely.

62. H. Christie to A. G. Manzie, "Treasury Comments on the Post Office's Model.," December 27, 1972, FV 87/6, TNA.

63. E. E. Baker to A. G. Manzie, "Post Office's Economic Appraisal of Exchange Equipment," November 8, 1972, FV 87/6, TNA.

64. H. Christie to A. G. Manzie, "Treasury Comments on the Post Office's Model," December 27, 1972, FV 87/6, TNA.

65. Hennessy, Morrison, and Townsend, "Routine Punctuated by Orgies," 5.

66. "IRC: Enquiry into the Telecommunications Industry" (1967), TCB 712/35, BTA.

67. "CPRS Summary of the Post Office's Modernisation Plan" (August 1972), FV 87/5, TNA.

68. "PO Model for Evaluating Alternative Equipment Strategies" (November 7, 1972), FV 87/6, TNA.

69. For a contemporaneous assessment of Concorde and the reactor's financial difficulties, see Henderson, "Two British Errors"; and for a historical overview, see Kelsey, "Picking Losers."

70. "Economic Appraisal of Exchange Equipment Strategies by Computer Model" (1972), FV 87/2, TNA.

71. For more on the Lighthill Report, see Agar, "What Is Science For?"

72. William Ryland to John Eden, "Exchange Equipment Strategy," February 1973, FV 87/3, TNA.

73. "Modernisation of Telephone Exchanges: Memorandum by the Minister of Posts and Telecommunications for the Cabinet Ministerial Committee on Economic Policy" (April 9, 1973), FV 87/4, TNA.

74. "Modernisation of Telephone Exchanges: Memorandum by the Minister of Posts and Telecommunications for the Cabinet Ministerial Committee on Economic Policy" (April 9, 1973), FV 87/4, TNA.

75. "Professor Merriman Retires . . . and so Does Dollis Hill."

76. Harris, *Automatic Switching in the UK*, 18.

77. "Post Office Board Meeting" (September 25, 1972), TCC 15/6, BTA; Harper, *Monopoly and Competition*, 110.

78. "System X" (September 1975), TCC 55/7/150, BTA.

79. Harris, *Automatic Switching in the UK*, 23.

80. Campbell-Kelly et al., *Computer*, 127–130.

81. "Development of the Trunk Network" (June 1973), TCC 55/5/96, BTA.

82. "Post Office Management Board Meeting" (July 2, 1973), TCC 15/8, BTA.

83. "Post Office/Industry Relationships in Telecommunications System Design" (September 1973), TCC 55/5/131, BTA; Harris, *Automatic Switching in the UK*, 26–27.

84. "Cable Procurement" (April 1975), TCC 55/7/50, BTA.

85. "System X" (September 1975), TCC 55/7/150, BTA.

86. Post Office Review Committee, *Report of the Post Office Review Committee*, 107–108; Hills, *Deregulating Telecoms*, 90–91.

87. Post Office Review Committee, *Report of the Post Office Review Committee*, 120.

88. Harris, *Automatic Switching in the UK*, 26–27.

89. "System X Arrives Early," 16.

90. Keith Joseph to Margaret Thatcher, "Telecommunications Monopoly," July 1, 1980, PREM 19/333, TNA.

91. D. H. Pitcher to John Hoskyns, "Post Office Monopoly," November 19, 1980, PREM 19/562, TNA.

92. Arnold Weinstock to Keith Joseph, "Post Office Switching Systems and Liberalisation," March 16, 1981, PREM 19/875, TNA.

93. J. S. Whyte to J. Samson, W. D. Morton, and D. H. Pitcher, "Organisation and Prospects for System X," May 18, 1982, PREM 19/876, TNA.

94. Kenneth Baker to Margaret Thatcher, "System X," September 29, 1982, PREM 19/877, TNA.

95. "£100m Boost for Network," 4.

96. Cawson et al., *Hostile Brothers*, 98–114.

97. Schiller, *Information and the Crisis Economy*, 2–3.

98. Schiller, *Information and the Crisis Economy*, 2.

99. Home Office, *Report of the Inquiry into Cable Expansion*; "Full Speed Ahead for Cable TV—Brittan."

100. Baker, "I. Information Technology: Industrial and Employment Opportunities."

101. *Metro—A British Car.*

102. Post Office, *Telephone Service and the Customer.*

103. "System X: The Key to Our Future" (1981), TCB 325/EHA 2753, BTA.

104. "System X: The Key to Our Future" (1981), TCB 325/EHA 2753, BTA.

105. "System X: The Way Ahead" (1982), TCB 325/EHA 2772, BTA.

106. "System X: The Key to Our Future" (1981), TCB 325/EHA 2753, BTA.

107. "Telecommunications Board: Network Modernisation and System X" (1980), TCC 62/1/64, BTA.

108. Clark et al., *Process of Technological Change.*

109. Clark et al., *Process of Technological Change*, 157.

110. Clark et al., *Process of Technological Change*, 195.

111. "System X: The Complete Approach" (1978), TCB 318/PH 2380, BTA; "System X: The Modernisation of British Telecom" (1984), TCB 325/EHA 2770, BTA.

112. Cookson, "BT to Computerize Directory Inquiries."

113. "Long Range Studies Report 31: Operator Services: The Scope for Further Automation in the 1980s and Later" (1972), TCC 252/31, BTA.

114. Agar, *Government Machine*, 298.

115. "New Services from System X" (1982), TCB 318/PH 3210, BTA.

116. Fagan, "Disconnecting Staff."

117. Husband and Pattinson, "BT Is Pulling the Plug."

118. Whitfield and Fagan, "BT Operators to Hold Strike Vote."

119. Gribben, "50,000 at BT Interested in Redundancy."

120. Levy, "Why 20,000 People Walked Out."

121. Ensmenger, *Computer Boys Take Over*; Abbate, *Recoding Gender*.

122. Godsmark, "BT Plans Radical Internal Shake-Up."

123. Hills, *Deregulating Telecoms*, 87.

124. Merriman, "Men, Circuits and Systems," 241.

CHAPTER 4

1. McGregor, "How Thatcher Killed the UK's Superfast Broadband."

2. Cochrane, "Home / Personal / About / CV."

3. Edge, "Technological Metaphor and Social Control."

4. Mayr, *Authority, Liberty, and Automatic Machinery*; Agar, *Government Machine*; Barry, *Political Machines*.

5. Morus, "'Nervous System of Britain'"; Otis, "Metaphoric Circuit."

6. MacDougall, *People's Network*, 249.

7. Sawhney, "Information Superhighway"; Stefik, *Internet Dreams*.

8. Sawhney, "Information Superhighway"; Thomas and Wyatt, "Shaping Cyberspace"; Wyatt, "Danger! Metaphors at Work."

9. Wise, *Values of Precision*; Schmidt and Werle, *Coordinating Technology*; Bowker and Star, *Sorting Things Out*; Barry, *Political Machines*; Gooday and Sumner, *By Whose Standards?*; Russell, *Open Standards and the Digital Age*; Yates and Murphy, *Engineering Rules*.

10. DeNardis, *Protocol Politics*.

11. Wise, *Values of Precision*; Yates and Murphy, *Engineering Rules*.

12. Abbate, *Inventing the Internet*, 147–180; Russell, *Open Standards and the Digital Age*, 161–228; Russell and Schafer, "In the Shadow of ARPANET."

13. Huurdeman, *Worldwide History of Telecommunications*, 337–343.

14. Fickers and Lommers, "Eventing Europe."

15. Jones and Edwards, "Post Office Network of Radio-Relay Stations. Part 1," 147.

16. Jones and Edwards, "Post Office Network of Radio-Relay Stations. Part 1."

17. Evelyn Sharp to Ernest Marples, "Post Office Tower," July 24, 1959, POST 122/1172, BTA.

18. Ernest Marples to Evelyn Sharp, "Post Office Tower," 1959, POST 122/1172, BTA.

19. "The Post Office Tower London" (1967), HIC W04/06 Buildings/BT Tower/Press cuttings and images, BTA.

20. Bowler, *Science for All*, 272.

21. "Dan Dare, Pilot of the Future"; "London's New Landmark."

22. Goldie, "'Radio Campanile.'"

23. "Post Office Tower Opening Ceremony Brochure" (1962), HIC W04/06 Buildings/BT Tower/Press cuttings and images, BTA; "The Post Office Tower Information Booklet" (1965), HIC W04/06 Buildings/BT Tower/Press cuttings and images, BTA.

24. "GPO Press Release: Full Circle" (1962), HIC W04/06 Buildings/BT Tower/ Construction, BTA.

25. Coase, "Lighthouse in Economics."

26. Ward, "Heart That Makes the Tower Tick."

27. "Birmingham Radio Tower" (1981), HIC W04/06 Buildings/BT Tower/Construction, BTA; "Progress: Birmingham Radio Tower" (1965), TCB 420/IRP (PR) 6, BTA.

28. Jones and Edwards, "Post Office Tower, London."

29. *Telecommunication Services for the 1990s*; "Telecommunications Services for the 1990s."

30. "British Telecom News Release: London Telecom Tower" (August 3, 1982), HIC W04/06 Buildings/BT Tower/Press cuttings and images, BTA.

31. "It's Lonely Up There"; Clark, "Tower to the People."

32. "Long-Distance Transmission by Waveguide."

33. Hecht, *City of Light*, 87, 172.

34. Hecht, *City of Light*, 87.

35. "A Year of Challenge."

36. Merlo, "Millimetric Waveguide System"; Hecht, *City of Light*, 174.

37. Richards, "Phone Trouble"; "Telephone Delays Up."

38. Breary, "A Long-Term Study of the United Kingdom Trunk Network, Part 1"; "UK Trunk Task Force Final Report, Volume 1: Guide and Summary Chapters" (1971), TCC 145/7, BTA.

39. Breary, "A Long-Term Study of the United Kingdom Trunk Network, Part 1," 213.

40. "Managing Director's Committee: Telecommunications" (November 17, 1972), TCC 55/4/194, BTA.

41. "Post Office Develops New Ways of Keeping in Touch" (1975), HIC W04/01 Network/Transmission Systems/Waveguides, BTA; "British Waveguide Plan—With Export Potential" (November 8, 1976), HIC W04/01 Network/Transmission Systems/ Waveguides, BTA.

42. "Millimetric Waveguide System Conference: Opening Address, J.H.H. Merriman" (November 9, 1976), TCC 711/31, BTA.

43. "Millimetric Waveguide System Conference: Opening Address, J.H.H. Merriman" (November 9, 1976), TCC 711/31, BTA.

44. "Post Office Board: Bristol-Reading Waveguide System" (1977), TCC 55/9/24, BTA.

45. "Post Office Management Board" (February 21, 1977), TCC 711/31, BTA.

46. "Bristol-Reading Waveguide System: Paper by Director of Network Planning" (1977), TCC 711/31, BTA.

47. Hale, *Wired Style*, 71.

48. Marchand, *Creating the Corporate Soul*, 60–61.

49. White, "Waveguides—Highways of Communication."

50. "British Waveguide On Show to World: A Successful Field Trial by Post Office" (November 8, 1976), HIC W04/01 Network/Transmission Systems/Waveguides, BTA; "British Waveguide Plan—With Export Potential" (November 8, 1976), HIC W04/01 Network/Transmission Systems/Waveguides, BTA; "British Waveguide Will Be World's First" (1977), TCC 711/31, BTA.

51. Sawhney, "Information Superhighway," 296.

52. Kaijser, "Helping Hand."

53. Habara, "ISDN."

54. The International Telegraph and Telephone Consultative Committee, *Fifth Plenary Assembly: Green Book*, III-3: Line Transmission:822.

55. Abbate, *Inventing the Internet*, 147–179; Russell and Schafer, "In the Shadow of ARPANET"; Russell, *Open Standards and the Digital Age*, 177–193.

56. Russell, *Open Standards and the Digital Age*, 197–228.

57. The International Telegraph and Telephone Consultative Committee, *Fifth Plenary Assembly: Green Book*, III-3: Line Transmission:834.

58. "Interactions Between Data Processing and Telecommunications" (June 1971), TCC 55/3/76, BTA; "Post Office Management Board Meeting" (July 2, 1973), TCC 15/8, BTA.

59. "Post Office Management Board Meeting" (March 15, 1976), TCC 15/17, BTA; see Russell, *Open Standards and the Digital Age*, 176, for more information about EPSS and similar schemes.

60. "Managing Directors' Committee: Telecommunications" (December 2, 1977), TCC 55/9/179, BTA.

61. Noam, *Telecommunications in Europe*, 360–361.

62. Hills, *Deregulating Telecoms*, 3.

63. Hills, *Deregulating Telecoms*, 3; Davids, "Privatisation and Liberalisation."

64. Hills, *Deregulating Telecoms*.

65. "X-Stream" (1981), TCB 318/PH 3096, BTA.

66. "X-Stream: A New Range of Digital Services from British Telecom" (1981), Folder 4, Box 80, MS2137: Joseph V. Charyk Papers, George Washington University Special Collections (henceforth GWSC).

67. "X-Stream" (1984), TCB 318/PH 3350, BTA.

68. "The Digital World of British Telecom" (1982), Folder 4, Box 80, MS2137, Joseph V. Charyk Papers, GWSC; "X-Stream" (1984), TCB 318/PH 3350, BTA; "ISDN: Martlesham '84 Stand Guide" (1984), TCB 318/PH 3625, BTA.

69. Negrine, "Cable Television in Great Britain," 104–107.

70. "Post Office Board Meeting" (April 24, 1967), TCB 14/5, BTA; "Rationalisation of the Local Distribution Network" (October 1968), TCB 54/2/45, BTA.

71. "Managing Director's Committee: Telecommunications" (November 1, 1974), TCC 55/6/178, BTA.

72. "Managing Director's Committee: Telecommunications" (November 1, 1974), TCC 55/6/178, BTA.

73. Committee of Inquiry on the Future of Broadcasting and Annan, *Report of the Committee on the Future of Broadcasting*; "British Telecommunications Board Meeting" (January 15, 1982), TCD 16/2, BTA.

74. "British Telecommunications Board Meeting" (January 15, 1982), TCD 16/2, BTA.

75. Information Technology Advisory Panel, *Report on Cable Systems*.

76. "British Telecommunications Board Meeting" (April 27, 1982), TCD 16/2, BTA.

77. "British Telecommunications Board Meeting" (October 26, 1982), TCD 16/2, BTA; Home Office, *Report of the Inquiry into Cable Expansion and Broadcasting Policy*.

78. "British Telecommunications Board Meeting" (November 23, 1982), TCD 16/2, BTA.

79. Home Office, *Development of Cable Systems and Services*.

80. Negrine, "Cable Television in Great Britain," 117.

81. Hecht, *City of Light*, 111.

82. Hecht, *City of Light*, 142–143.

83. Midwinter, "Optical-Fibre Transmission Systems."

84. "Optical Fibre Systems."

85. "Recent Major Events in the Evolution of British Telecom's Optical-Fibre Network."

86. "Telecom's Optical Triumph"; "World's First Commercial 140 Mbit/s Optical Link Fibre Tested."

87. "New Optical Fibre Record Set to Cut Costs" (November 21, 1985), HIC W04/03 Network/Cable/Fibre-Optic, BTA; "Queen's Award for Telecom Research" (April 17, 1985), HIC W04/03 Network/Cable/Fibre-Optic, BTA.

88. Hecht, City of Light, 232.

89. The electricity market was not liberalized and privatized in the UK until 1990.

90. Merriman, Driving Spaces; Cole, "About Britain: Driving the Landscape of Britain (at Speed?)."

91. Advisory Council on Science and Technology, Optoelectronics: Building on Our Investment; Communications Steering Group, Infrastructure for Tomorrow.

92. Communications Steering Group, Infrastructure for Tomorrow, 5.

93. Communications Steering Group, Infrastructure for Tomorrow, 5.

94. DeNardis, Protocol Politics.

95. "Longest Lightlines for UK Cities"; "On the Lightlines . . ."; "'Super' Link Installed"; "Lightlines" (1988), TCC 474/HF 83E, BTA.

96. Rowbotham, "Plans for a British Trial of Fibre."

97. Hecht, City of Light, 219–224.

98. Larger, "BT Halts Move."

99. Wise, "Warning Bells Ring for BT."

100. Department of Trade and Industry, Competition and Choice.

101. Rudge, "Why the DTI Is Out of Order."

102. Parker-Jervis, "Cable TV Firms Set"; Keegan, "Through a Glass Fibre Darkly"; Hutton, "Why Britain Is So Slow."

103. "Find Out How the Business Highway Can Help You Work Faster." (1998), TCE 306/PHME 33098, BTA; "Get on the BT Highway" (1998), TCE 306/PHME 32863, BTA.

104. "Highway 1" (1998), TCE 305/V 99576, BTA.

105. Hale, Wired Style, 71.

106. Sawhney, "Information Superhighway," 304–309.

107. Gore Jr., "Remarks at Superhighway Summit."

108. Thomas and Wyatt, "Shaping Cyberspace"; Wyatt, "Danger! Metaphors at Work."

109. Cawson et al., Hostile Brothers; Edgerton, Rise and Fall of the British Nation.

CHAPTER 5

1. Vallis, "Martlesham Heath Village"; "Redevelopment of Martlesham Heath, East Suffolk" (February 1965), TCB 391/1/3/9/1, BTA.

2. Saxenian, *Regional Advantage*; O'Mara, *Cities of Knowledge*; Lécuyer, *Making Silicon Valley*; O'Mara, *The Code*.

3. Ceruzzi, *Internet Alley*; Misa, *Digital State*.

4. Sagarena, "Building California's Past"; Malfona, "Building Silicon Valley."

5. Hecht, *Radiance of France*, 239.

6. Kargon and Molella, *Invented Edens*, 13.

7. Clapson, *Invincible Green Suburbs*; Meller, *Towns, Plans and Society*; Alexander, *Britain's New Towns*.

8. Wakeman, *Practicing Utopia*, 33.

9. Ortolano, *Thatcher's Progress*.

10. Wetherell, "Freedom Planned," 266.

11. Ward, "Consortium Developments Ltd and the Failure of 'New Country Towns'"; Hall, *Cities of Tomorrow*, 441–442.

12. Giddens, *Social Theory and Modern Sociology*, 153–165.

13. Harvey, *Condition of Postmodernity*, 85–87, 217.

14. Castells, *Rise of the Network Society*, 440–448, 464–465, 493.

15. "Feasibility of Existing Site" (1963), TH/FB/528, BTA; "Dollis Hill: Estimated Growth R. Branch 1964–74" (1964), TH/FB/362, BTA.

16. "Dispersal of Post Office Research Station" (April 30, 1964), TH/FB/528, BTA.

17. For a more comprehensive history of Dollis Hill's early years, see Haigh, "'To Strive, to Seek, to Find.'"

18. Barres-Baker, "Brief History of the London Borough of Brent."

19. "Middlesex XI.14 (Includes: Willesden)," 1914; "Middlesex XI.14 (Includes: Willesden)," 1938.

20. "Dispersal: Post Office Research Station, Dollis Hill" (March 16, 1964), TH/FB/528, BTA.

21. "Conclusions (60) 11: General Directorate Meeting" (June 27, 1960), POST 72/1083, BPMA.

22. Llewellyn, "Producing and Experiencing Harlow"; Childs, "Harlow."

23. "Dispersal: Post Office Research Station, Dollis Hill" (March 16, 1964), TH/FB/528, BTA.

24. "Dispersal of Government Work" (1963), MAF 229/22, TNA.

25. "Postmaster General: Dispersal: Dollis Hill Research Station" (May 1, 1964), TH/FB/528, BTA.

26. Phillips, "Black Minority Ethnic Concentration, Segregation and Dispersal in Britain"; Esteves, *"Desegregation" of English Schools*.

27. Light, *From Warfare to Welfare*; O'Mara, *Cities of Knowledge*.

28. Reginald Bevins to John Boyd-Carpenter, "Draft: Dispersal of Post Office Research Station," April 1964, TH/FB/528, BTA.

29. D. J. Kinder to C. H. Coates, "Dollis Hill Research Station," February 5, 1964, TH/FB/528, BTA; N. P. Lester to G. H. Metson, "Post Office Research Department," February 12, 1964, TH/FB/528, BTA.

30. "Postmaster General: Dispersal: Dollis Hill Research Station" (March 6, 1964), TH/FB/528, BTA.

31. "HC Deb: Ministry (Staff)" (January 28, 1964), Vol 688 c34W, Hansard; "HC Deb: Government Offices (Dispersal)" (January 29, 1964), Vol 677 cc486–506, Hansard; "Note by J.R. Bevins, Postmaster General" (1964), TH/FB/528, BTA.

32. G. H. Metson to D. J. Kinder, "East Suffolk County Council," February 7, 1964, TH/FB/528, BTA; T. B. Oxenbury to G. H. Metson, "Projected Move from London," February 4, 1964, TH/FB/528, BTA.

33. "Deputy Director General: Dispersal: Dollis Hill Research Station" (March 16, 1964), TH/FB/528, BTA; "Notes of a Meeting Held at the House of Commons" (December 19, 1962), TH/FB/528, BTA.

34. "Dispersal of Dollis Hill Research Station" (March 18, 1964), TH/FB/528, BTA.

35. "Dispersal of Dollis Hill Research Station." (March 18, 1964), TH/FB/528, BTA.

36. "The South-East Study" (1964), TH/FB/528, BTA.

37. "Dispersal of Dollis Hill Research Station" (March 18, 1964), TH/FB/528, BTA.

38. "Note to Director General from A.H. Mumford" (June 22, 1964), TH/FB/528, BTA; "HC Deb: Research Centre, Dollis Hill" (July 28, 1964), Vol 699 cc234–5W, Hansard.

39. "Draft Press Statement by County Planning Officer: Development at Martlesham for the Post Office Engineering Research Station" (July 22, 1964), TH/FB/528, BTA.

40. Beloff, *Plateglass Universities*; Muthesius, *Post-War University*, 104–105.

41. Muthesius, *Post-War University*, 107.

42. Muthesius, *Post-War University*, 149.

43. "University of Essex: Chair in Telecommunication Systems"; "University of Essex: Electronic Engineering."

44. "Phone Foresight"; "University of Essex/Post Office Research Centre. SRC Case Studentship: Optical Communications."

45. "University of Essex Celebrates Four Decades."

46. "Laboratories Visited in USA" (November 1964), TCB 391/2/4, BTA.

47. "Laboratories Visited in USA" (November 1964), TCB 391/2/4, BTA.

48. Knowles and Leslie, "Industrial Versailles."

49. "GPO/MPBW Meeting: Move of the Research Station" (August 7, 1967), TCB 391/2/2, BTA.

50. "Buildings and Welfare Department Meeting: Martlesham Heath Research Centre" (1966), TCB 391/2/6, BTA.

51. Muthesius, *Post-War University*, 107–174; Knowles and Leslie, "Industrial Versailles," 26.

52. Light, *From Warfare to Welfare*.

53. Floyd, "Design of Martlesham Research Centre: Part 1"; Floyd, "Design of Martlesham Research Centre: Part 2."

54. Floyd, "Design of Martlesham Research Centre: Part 1," 147.

55. Floyd, "Design of Martlesham Research Centre: Part 2," 263–264.

56. West, "Facilities for Experimental Work."

57. Godin and Schauz, "Changing Identity of Research."

58. Godin, "Research and Development."

59. Knowles and Leslie, "Industrial Versailles."

60. "R&D: New Horizons in Telecommunications" (1975), HIC W04/06 Business Ops/R&D/Dollis Hill—Martlesham Heath—The Circuit Laboratory, BTA.

61. "Redevelopment of Martlesham Heath, East Suffolk" (February 1965), TCB 391/1/3/9/1, BTA.

62. "Redevelopment of Martlesham Heath, East Suffolk" (February 1965), TCB 391/1/3/9/1, BTA, 13.

63. "Redevelopment of Martlesham Heath, East Suffolk" (February 1965), TCB 391/1/3/9/1, BTA, 4.

64. "Martlesham Heath Township Proposals."

65. "Redevelopment of Martlesham Heath, East Suffolk" (February 1965), i, TCB 391/1/3/9/1, BTA.

66. "Redevelopment of Martlesham Heath, East Suffolk" (February 1965), i, TCB 391/1/3/9/1, BTA, 20.

67. "Martlesham Move Committee, Progress Report 6" (July 1969), TCB 391/1/4/1, BTA; "Martlesham Move Committee, Progress Report 11" (October 1971), TCB 391/1/4/1, BTA.

68. "Martlesham Move Committee, Progress Report 10" (May 1971), TCB 391/1/4/1, BTA.

69. Vallis, "Martlesham Heath Village"; Parker and Darley, "Martlesham Heath Village," 492; Aldous, "Controlled Chaos Proves a Winner."

70. "Twentieth Century Village."

71. Parker and Darley, "Martlesham Heath Village," 485.

72. Vallis, "Martlesham Heath Village."

73. Aldous, "Controlled Chaos Proves a Winner."

74. "Building Dossier: Martlesham Heath"; Prichard, "Village Values."

75. Aldous, "Suffolk Sensibilities," 15; "Building Dossier: Martlesham Heath," 48.

76. "Building Dossier: Martlesham Heath," 49.

77. Kargon and Molella, *Invented Edens*, 149.

78. Jencks, *New Paradigm in Architecture*, 67; Samuel, *Theatres of Memory*, 59–63.

79. Prichard, "Village Values"; Rybczynski, "Behind the Façade"; Meades, *Museum without Walls*, 367.

80. Harvey, *The Condition of Postmodernity*, 40, 75.

81. Prichard, "Village Values," 43.

82. Aldous, "Controlled Chaos Proves a Winner."

83. "British Telecommunications Board Meeting" (December 22, 1981), TCD 16/1, BTA.

84. "BT Enterprise in Venture."

85. "Martlesham Enterprises" (1981), TCD 69/1/35, BTA; McNally, *Corporate Venture Capital*, 61.

86. Moreton, "Survey: Science Parks."

87. Cane, "Hewlett-Packard Purchase."

88. "BT Exits Tech Mahindra."

89. "British Acquire 20% of MCI"; Misa, *Digital State*, 134.

90. "AT&T, British Telecom Set $10B Venture."

91. "Corning in Deal with NetOptix."

92. "First Chapter: Celebrating 30 Years of BT Laboratories" (August 1999), HIC W04/06 Business Ops/R&D/Dollis Hill—Martlesham Heath—The Circuit Laboratory, BTA.

93. "Venturing to a Higher Plane: BT R&D Innovation Report" (2000), HIC W04/06 Business Ops/R&D/Dollis Hill—Martlesham Heath—The Circuit Laboratory, BTA.

94. "Venturing to a Higher Plane: BT R&D Innovation Report" (2000), HIC W04/06 Business Ops/R&D/Dollis Hill—Martlesham Heath—The Circuit Laboratory, BTA; Fletcher and Fagan, "BT Sets Up £2bn Venture Arm."

95. "Venturing to a Higher Plane: BT R&D Innovation Report" (2000), HIC W04/06 Business Ops/R&D/Dollis Hill—Martlesham Heath—The Circuit Laboratory, BTA.

96. "Companies | Innovation Martlesham."

97. "UK Ministers Defend Chinese Deals"; Yueh, "Huawei Boss Says US Ban 'Not Very Important'"; "Huawei: UK Bans New 5G Network Equipment."

98. O'Mara, *Cities of Knowledge*.

99. Campbell-Kelly, *ICL*.

100. Barrow, "BT's Bunker Is a Brave New World"; Wilby, "Inside Adastral."

101. "Martlesham Heath: Home of Experimental Units" (January 1967), TCB 391/1/3/9/1, BTA.

102. "First Chapter: Celebrating 30 Years of BT Laboratories" (August 1999), HIC W04/06 Business Ops/R&D/Dollis Hill—Martlesham Heath—The Circuit Laboratory, BTA.

103. "First Chapter: Celebrating 30 Years of BT Laboratories" (August 1999), HIC W04/06 Business Ops/R&D/Dollis Hill—Martlesham Heath—The Circuit Laboratory, BTA.

104. "Tommy Flowers Institute: BT Adastral Park."

105. "Martlesham Heath: 100 Groundbreaking Years."

106. "Main Event—Martlesham Heath: 100 Groundbreaking Years."

107. "Martlesham Heath: 100 Groundbreaking Years."

108. Samuel, *Theatres of Memory*, 65, 146–174; Wright, *On Living in an Old Country*, 6.

109. "No Adastral New Town."

110. "No Adastral New Town."

111. Mclaughlin, "Have Your Say."

112. Cahan, "Geopolitics and Architectural Design"; Agar, *Science and Spectacle*.

113. See Ortolano, *Thatcher's Progress*, 108–142 for the origins of this quote, and yet Ortolano argues that welfare state modernism's demise was engineered by private developers, rather than rejected by the public.

114. Dennis, "Accounting for Research"; Knowles and Leslie, "Industrial Versailles."

CHAPTER 6

1. *British Telecom International . . . It's You We Answer To.*

2. *AT&T's Universal Telephone Service.*

3. Kennedy, "Imperial Cable Communications and Strategy; Headrick, *Invisible Weapon*; Hills, *Telecommunications and Empire*; Finn and Yang, *Communications Under the Seas.*

4. Müller, *Wiring the World.*

5. Tworek, *News from Germany*; Tworek, "How Not to Build a World Wireless Network."

6. Butrica, *Beyond the Ionosphere*; Slotten, "Satellite Communications, Globalization, and the Cold War"; Parks and Schwoch, *Down to Earth*; Parks, *Cultures in Orbit*; Schwoch, *Global TV.*

7. Müller, *Wiring the World*; Müller and Tworek, "'Telegraph and the Bank'"; Schwoch, *Global TV*; Tworek, "How Not to Build a World Wireless Network"; Collins, "One World . . . One Telephone"; Collins, *Telephone for the World.*

8. Fari, "Introduction."

9. Balbi and Fickers, *History of the International Telecommunication Union*; Schot and Lagendijk, "Technocratic Internationalism in the Interwar Years"; Henrich-Franke, "Comparing Cultures of Expert Regulation."

10. Slotten, "Satellite Communications, Globalization, and the Cold War"; Slotten, "International Governance, Organizational Standards."

11. Hills, *Deregulating Telecoms*; Hulsink, *Privatisation and Liberalisation in European Telecommunications*; Thatcher, *Internationalisation and Economic Institutions*; Lipartito, "Regulation Reconsidered."

12. Barry, *Political Machines.*

13. Barry, *Political Machines*, 40.

14. Stine and Tarr, "At the Intersection of Histories"; Agar and Ward, *Histories of Technology, the Environment and Modern Britain* Tully, "Victorian Ecological Disaster"; Starosielski, *Undersea Network*; Schwoch, *Wired into Nature*; Ward, "Oceanscapes and Spacescapes"; Rozwadowski, "Ocean's Depths"; Launius, "Writing the History of Space's Extreme Environment"; Rand, "Orbital Decay."

15. Gleichmann et al., "Repeater Design."

16. Kelly et al., "Transatlantic Telephone Cable"; Elmendorf and Heezen, "Oceanographic Information"; Snoke, "Resistance of Organic Materials and Cable Structures"; Jack, Leech, and Lewis, "Route Selection and Cable Laying for the Transatlantic Cable System."

17. AT&T and N. W. Ayer, "The Sea Could Make A 'Meal' of Telephone Cables" (1958), Folder 1, Box 32, Series 3, Collection 59: N. W. Ayer Advertising Agency Records, National Museum of American History Archive Center (henceforth NMAH); AT&T and N. W. Ayer, "Why We Have Our Own 'Ocean'" (1959), Folder 1, Box 32, Series 3, Collection 59: N. W. Ayer Advertising Agency Records, NMAH.

18. AT&T and N. W. Ayer, "Beneath the Broad Atlantic" (1954), Folder 3, Box 13, Series 3, Collection 59: N. W. Ayer Advertising Agency Records, NMAH; AT&T and N. W. Ayer, "You Can Telephone Britain over New Stormproof Cables" (1956), Folder 2, Box 24, Series 3, Collection 59: N. W. Ayer Advertising Agency Records, NMAH; AT&T and N. W. Ayer, "Transatlantic Telephone Cable Is Now Being Laid" (1955), Folder 3, Box 13, Series 3, Collection 59: N. W. Ayer Advertising Agency Records, NMAH.

19. AT&T and N. W. Ayer, "Tele-Facts: Laying the First Atlantic Telephone Cable" (1955), Folder 2, Box 32, Series 3, Collection 59: N. W. Ayer Advertising Agency Records, NMAH.

20. Craig, "Equipping Ourselves for Today's Responsibilities"; Duncan, "Communications and Defense."

21. "Route of the Proposed Transatlantic Cable" (July 1953), DO 35/4940, TNA.

22. Ben Barnett to Alexander Little, "Secret: From Washington to Foreign Office," April 11, 1953, FO 371/105731, TNA; R. R. G. Watts to Gibson, "Phone Call with Wolverson and Barnett on the Trans-Atlantic Telephone Scheme," June 30, 1953, DO 35/4940, TNA.

23. "Route of the Proposed Transatlantic Cable" (July 1953), DO 35/4940, TNA.

24. J. J. S. Garner to Ben Barnett, "Canadian High Commissioner," July 24, 1953, DO 35/4940, TNA.

25. Ben Barnett to Alexander Little, "Secret: From Washington to Foreign Office," April 11, 1953, FO 371/105731, TNA; Kelly et al., "A Transatlantic Telephone Cable," 128.

26. Cable to the Continent.

27. Krige, American Hegemony.

28. "Speeches at the Opening of the Transatlantic Telephone Cable" (1956), HIC W04/03 Network/Cable/Submarine Cables/Trans-Atlantic, BTA.

29. Sir Gordon Radley, "The Transatlantic Telephone Cable" (September 27, 1956), HIC W04/03 Network/Cable/Submarine Cables/Trans-Atlantic, BTA.

30. Kipling, "Deep Sea Cables."

31. Hills, "Regulation, Innovation and Market Structure."

32. Crawford et al., "Research Background of the Telstar Experiment"; "Project Telstar" (1962), HIC W04/02 Network/Transmission Systems/Satellites 2, BTA.

33. Crawford et al., "Research Background of the Telstar Experiment"; "Project Telstar" (1962), HIC W04/02 Network/Transmission Systems/Satellites 2, BTA.

34. AT&T and N. W. Ayer, "From Beyond the Sky to Beneath the Seas" (1958), Folder 3, Box 15, Series 3, Collection 59: N. W. Ayer Advertising Agency Records, NMAH.

35. AT&T and N. W. Ayer, "Between Outer Space And The Deep Sea There's A Wide Range Of Opportunity In The Bell Telephone Companies" (1964), Folder 1, Box 30, Series 3, Collection 59: N. W. Ayer Advertising Agency Records, NMAH; AT&T and N. W. Ayer, "Progress in the Bell System . . ." (1964), Folder 1, Box 30, Series 3, Collection 59: N. W. Ayer Advertising Agency Records, NMAH.

36. "Facts from Space via Telstar."

37. "Van Allen Sees Science 'Clique'"; "AT&T Sets Telstar II Launch for Spring."

38. Rand, "Orbital Decay," 127; Higuchi, "Atmospheric Nuclear Weapons Testing," 319; Hamblin, *Arming Mother Nature*, 12.

39. McNeill and Unger, "Introduction: The Big Picture," 11.

40. Raymond, "U.S. Gives Soviet Report on Search of Fishing Vessel."

41. Lovell, "Challenge of Space Research"; "Protests Continue Abroad."

42. "More Needles in Space."

43. For a study from the perspective of the ground station, see Waff, "Project Echo, Goldstone, and Holmdel."

44. Agar, "Making a Meal of the Big Dish."

45. "Ten Years of Technological Progress" (1972), HIC W04/02 Network/Transmission Systems/Satellites 2, BTA.

46. "Progress: Goonhilly" (1962), TCB 420/IRP (PR) 1, BTA.

47. "Prime Minister: TELSTAR" (July 1962), TCB 2/184, BTA.

48. Agar, *Science and Spectacle*; Cahan, "Geopolitics and Architectural Design."

49. "Prime Minister: TELSTAR" (July 1962), TCB 2/184, BTA.

50. D. I. Dalgleish and V. C. Meller, "Communication-Satellite Earth Stations: The First Ten Years" (1973), HIC W04/02 Network/Transmission Systems/Satellites 2, BTA; Edward Fennessy, "Provision of Fourth UK INTELSAT Earth Terminal: Development of New Earth Station Site" (October 1974), TCC 55/6/145, BTA.

51. Weppler, "Radio Spectrum Squeeze."

52. Slotten, "Satellite Communications, Globalization, and the Cold War"; Slotten, "International Telecommunications Union."

53. Wright, "Formulation of British and European Policy Toward an International Satellite Telecommunications System: The Role of the British Foreign Office"; Slotten, "International Governance, Organizational Standards."

54. "Draft Paper for Circulation to the Members of the Conference of European Postal and Telecommunications Administrations: Commonwealth Conference on Satellite Communications" (July 25, 1962), TCB 2/182, BTA.

55. Slotten, "International Telecommunications Union."

56. "Experimental Earth Station at Goonhilly Downs" (Geneva, 1963), Document 76, EARC-63, ITU Conferences Collection, International Telecommunications Union Archive (henceforth ITU); "Draft Resolution—The Interconnection of Communication-Satellite Systems and Other Transmission Systems" (Geneva, 1963), Document 121, EARC-63, ITU Conferences Collection, ITU.

57. "Goonhilly Computer Helps International Planning" (October 28, 1963), TCB 2/179, BTA.

58. Slotten, "International Governance, Organizational Standards," 538.

59. McLuhan, *Gutenberg Galaxy*; Turner, *From Counterculture to Cyberculture*.

60. "The Communications Explosion."

61. "Early Bird Advert by Hughes, Aviation Weekly" (May 17, 1965), OE-037000–01, National Air and Space Museum Archive Center (henceforth NASM).

62. "Fact Sheet II: Early Bird and Future Communications Satellites, COMSAT" (1965), OE-037000–03, NASM.

63. "Johnson, Sato, Exchange Messages via Pacific Satellite, COMSAT" (1967), Folder 3, Box 58, Acc. No. 1987–0125, NASM.

64. "INTELSAT Agreements Signed, U.S. Department of State" (August 20, 1971), OI-435000–01, NASM.

65. von Braun, "Now At Your Service," 56.

66. TRW, "INTELSAT III Press Handbook" (1969), OI-435100–01, NASM; Clarke, "Spinoff from Space," 30.

67. Schwoch, *Global TV*; Parks, *Cultures in Orbit*.

68. COMSAT, "New Communications Era" (1967), Folder 5, Box 58, Acc. No. 1987–0125, NASM; von Braun, "Now at Your Service—The World's Most Talkative Satellite," 56.

69. COMSAT, "Early Bird to Provide Emergency Service after Transatlantic Cable Break" (June 18, 1965), OE-037000–03, NASM.

70. COMSAT, "Interruption of Service on Two Transatlantic Cables" (February 15, 1968), OI-435000–01, NASM.

71. "Commercial Satellite Communications, INTELSAT." (February 17, 1969), Folder 9, Box 58, Acc. No. 1987–0125, NASM.

72. TRW, "INTELSAT III Press Handbook" (1969), OI-435100–01, NASM.

73. "First Intelsat 4 Placed in Orbit"; "Switchboard In Orbit."

74. "Long Distance Cable and Satellite Systems" (May 6, 1968), TCB 711/2/5, BTA.

75. "Joint Submarine Systems Development Unit (P.O. and C.&.W.): General Appraisal of Present and Future Activities" (March 3, 1967), TCB 711/29/9, BTA.

76. Bates, "Ploughing the Seabed."

77. "The State of the Art of Diving at the Present Time and in the Foreseeable Future, with Particular Reference to the Repair of Buried Submarine Cable" (1970), TCB 711/30/5, BTA.

78. I. R. Finlayson to F. A. Hough, "Proposed Organisation of R&D by the Underwater Engineering Group, CIRIA," November 19, 1969, TCB 711/30/5, BTA.

79. Edward Fennessy, "Submarine Recovery" (September 13, 1973), TCB 711/30/5, BTA; "Purchase of Submersible Vehicles for North Atlantic Cable Maintenance" (January 28, 1975), TCB 711/30/5, BTA.

80. The British Post Office, "Cut & Hold Grapnel" (1979), TCB 711/35/3, BTA; Cosier, "Getting to Grips with Undersea Cables."

81. "Report of the First Session of the North Atlantic Systems Conference" (February 18, 1977), Folder 11, Box 62, MS2137: Joseph V. Charyk Papers, GWSC.

82. J. Hodgson, "Conference on North Atlantic Systems: Discussions in New York with AT&T and CTNE, 8–9 July 1976" (July 12, 1976), TCB 711/29/25, BTA.

83. Jack Oslund to R. R. Colino, "Eastbourne Meeting," March 1, 1977, Folder 11, Box 62, MS2137: Joseph V. Charyk Papers, GWSC.

84. "Go-Ahead for TAT-8."

85. Smith, "BT Bites Back . . . New Cable More Than a Match for 'Jaws'"; "INTELSAT V Facts" (1981), OI-435400–01, NASM.

86. Hills, *Deregulating Telecoms*, 157–179.

87. Hills, *Deregulating Telecoms*, 179.

88. "X-Stream: A New Range of Digital Services from British Telecom" (1981), Folder 4, Box 80, MS2137: Joseph V. Charyk Papers, GWSC.

89. Hills, *Deregulating Telecoms*, 19.

90. Hills, *Deregulating Telecoms*, 170.

91. "The London Teleport" (October 1984), TCD 265/BTIPR 263, BTA.

92. "Uniting the Business World" (1984), TCD 265/BTIPR 253, BTA; "Videoconferencing" (1986), TCE 310/BTI AP 1174(a), BTA.

93. "The Information World" (1986), TCE 310/BTI AP(a), BTA.

94. AT&T and N. W. Ayer, "Data 99.5% Pure" (1982), Folder 3, Box 28, Series 4, Collection 59: N. W. Ayer Advertising Agency Records, NMAH; AT&T and N. W. Ayer,

"Call the UK $1.25" (1982), Folder 6, Box 28, Series 4, Collection 59: N. W. Ayer Advertising Agency Records, NMAH.

95. AT&T and N. W. Ayer, "Our Network Is the Foundation for the Information Age" (1983), Folder 1, Box 30, Series 4, Collection 59: N. W. Ayer Advertising Agency Records, NMAH.

96. AT&T and N. W. Ayer, "Issues of the Information Age: Promises Kept, Promises to Keep" (1986), Folder 2, Box 31, Series 4, Collection 59: N. W. Ayer Advertising Agency Records, NMAH.

97. Bohn, "Information Technology in Development"; Parker, "Information and Society"; Lamberton, "Information Revolution."

98. Slobodian, *Globalists*.

CHAPTER 7

1. Patrick Jenkin to Leon Brittan, "Future Policy on Telecommunications," March 12, 1982, PREM 19/875, TNA.

2. Margaret Thatcher, "Speech Opening Conference on Information Technology" (The Barbican Centre, London, December 8, 1982), 105067, Thatcher MSS (Digital Collection), http://www.margaretthatcher.org/document/105067.

3. John Redwood to Margaret Thatcher, "British Telecom Meeting," July 27, 1984, PREM 19/1345, TNA.

4. "British Telecom: Offer for Sale of Ordinary Shares" (1984), TCE 49/2, BTA; Newman, *Selling of British Telecom*, 215.

5. Barbrook and Cameron, "Californian Ideology"; Turner, *From Counterculture to Cyberculture*.

6. Dyson et al., "Cyberspace and the American Dream."

7. Turner, *From Counterculture to Cyberculture*, 232.

8. Edwards, "'Financial Consumerism'"; Edwards, "'Manufacturing Capitalists.'"

9. Kynaston, *City of London*; Roberts and Kynaston, *City State*; Michie, *City of London*; Wójcik, *Global Stock Market*.

10. Duménil and Lévy, *Capital Resurgent*; Oren and Blyth, "From Big Bang to Big Crash."

11. Bellringer and Michie, "Big Bang in the City of London."

12. Cassis, *Capitals of Capital*, 246; Capie, *Bank of England*, 780–783.

13. Davies, *City of London and Social Democracy*; Rollings, "Cracks in the Post-War Keynesian Settlement?"; Rollings, "Organised Business and the Rise of Neoliberalism"; Jackson, "The Think-Tank Archipelago."

14. Bijker, Hughes, and Pinch, *Social Construction of Technological Systems*; Kline and Pinch, "Users as Agents"; Oudshoorn and Pinch, *How Users Matter*.

15. Martin, "Communication and Social Forms"; Martin, *"Hello Central?"*; Fischer, *America Calling*; MacDougall, *People's Network*.

16. Scott, "Still a Niche Communications Medium."

17. Milne, "British Business and the Telephone"; Milne, "Business Districts, Office Culture," 199.

18. Hills, *Deregulating Telecoms*; Cawson et al., *Hostile Brothers*, 84, 357–369.

19. MacKenzie, *Material Markets*.

20. MacKenzie, "Opening the Black Boxes of Global Finance"; Knorr-Cetina, "From Pipes to Scopes"; MacKenzie et al., "Drilling through the Allegheny Mountains"; MacKenzie, "Material Political Economy"; MacKenzie, "Material Signals."

21. Cassis, *Capitals of Capital*; Müller and Tworek, "'Telegraph and the Bank'"; Preda, "Socio-Technical Agency in Financial Markets"; Preda, *Framing Finance*.

22. MacKenzie, "Mechanizing the Merc"; Kennedy, "Machine in the Market."

23. Michie, "Friend or Foe?"

24. Pardo-Guerra, *Automating Finance*; Sweetman, *Cyber and the City*.

25. For an exception, see Cassis, De Luca, and Florio, *Infrastructure Finance in Europe*.

26. Davies, *City of London and Social Democracy*, 140–180.

27. Cyril H. Kleinwort to John Stonehouse, "Lunch at Sandy Glen," September 18, 1968, 6A403/1, Bank of England Archive (henceforth BOE); "Committee on Invisible Exports: Telecommunications Sub-Committee" (June 16, 1969), 6A403/1, BOE.

28. Sweetman, *Cyber and the City*, 78.

29. Sweetman, *Cyber and the City*, 32–39.

30. "Committee on Invisible Exports: Telecommunications Sub-Committee" (June 16, 1969), 6A403/1, BOE; "Committee on Invisible Exports: Telecommunications Sub-Committee" (February 17, 1970), 6A403/1, BOE.

31. "Memorandum: G.P.O. Telex" (December 23, 1968), 6A403/1, BOE.

32. "Present Shortcomings in International Telecommunications" (June 26, 1973), 6A403/1, BOE.

33. Sweetman, *Cyber and the City*, 94–95.

34. "City Telecommunications Group Meeting: International Telecommunications" (October 8, 1973), 6A403/1, BOE; "City Telecommunications Group Meeting" (December 19, 1973), 6A403/1, BOE.

35. "City Telecommunications Group Meeting with the Post Office" (January 31, 1974), 6A403/1, BOE.

36. "Meeting Minutes: London as a World Financial Centre" (April 4, 1974), 6A403/1, BOE.

37. W. M. Clarke, "City Telecommunications Committee" (1974), 6A403/2, BOE.

38. Kennedy, "Machine in the Market"; Michie, "Friend or Foe?"

39. Francis Sandilands to William Barlow, "Post Office Telecommunications," December 9, 1977, 6A403/2, BOE; William Barlow to Francis Sandilands, "Arrangements for Contracts Relating to Wide Band Circuits," January 9, 1978, 6A403/2, BOE.

40. Campbell-Smith, *Masters of the Post*, 550–551.

41. "Communications Services in the United Kingdom for Business Users" (August 1979), 6A403/3, BOE.

42. "Association of Telecommunications Users: Inaugural Meeting" (October 17, 1979), 6A403/3, BOE.

43. "Telecommunications in the City" (September 1980), 6A403/3, BOE.

44. Parker, *Official History of Privatisation*, 1: The Formative Years, 1970–1987:291.

45. Barclays Merchant Bank Ltd., "Proposals for the Issue of Shares by British Telecom: A New Strategy for Widening Share Ownership in the United Kingdom" (February 2, 1984), PREM 19/1345, TNA.

46. John Redwood to Margaret Thatcher, "British Telecom Meeting," July 27, 1984, PREM 19/1345, TNA.

47. Parker, *Official History of Privatisation*, 1: The Formative Years, 1970–1987:300–302.

48. "British Telecom: Offer for Sale of Ordinary Shares" (1984), TCE 49/2, BTA; Parker, *Official History of Privatisation*, 1: The Formative Years, 1970–1987:301.

49. Parker, *Official History of Privatisation*, 1: The Formative Years, 1970–1987:307.

50. "British Telecom" (November 16, 1984), HC Deb Vol 67 cc915–922, Hansard.

51. Parker, *Official History of Privatisation*, 1: The Formative Years, 1970–1987:291–316.

52. Parker, *Official History of Privatisation*, 1: The Formative Years, 1970–1987:291–316.

53. Baker, *Turbulent Years*, 84; Margaret Thatcher, "The Principles of Thatcherism" (Seoul, South Korea, September 3, 1992), 108302, Thatcher MSS (Digital Collection), http://www.margaretthatcher.org/document/108302.

54. "British Telecom Share Offer" (1984), PREM 19/1599, TNA.

55. "Telecommunications in the City of London: A Note on the Telecity Studies" (October 23, 1980), 6A403/3, BOE.

56. "Lightlines for the Heart of London" (October 24, 1983), HIC W04/03 Network/Cable/Fibre-Optic, BTA.

57. "The London Teleport" (October 1984), TCD 265/BTIPR 263, BTA.

58. Sweetman, *Cyber and the City*, 68–69, 168–169.

59. "Telecommunications in the City of London: A Note on the Telecity Studies" (October 23, 1980), 6A403/3, BOE.

60. Edwards, "'Manufacturing Capitalists'"; "Wider Share Ownership Policy Group" (April 1968), CRD 4/4/176, Conservative Party Archive.

61. George Jefferson to Patrick Jenkin, "Privatisation," April 19, 1982, TCD 69/2/45, BTA.

62. Lean, *Electronic Dreams*, 92.

63. Kline, *Cybernetics Moment*, 203.

64. Patrick Jenkin to Leon Brittan, "Future Policy on Telecommunications," March 12, 1982, PREM 19/875, TNA.

65. Cawson et al., *Hostile Brothers*, 92–93.

66. Hills, *Deregulating Telecoms*, 93.

67. Oakley and Owen, *Alvey*; Lean, *Electronic Dreams*.

68. Garvey, "Artificial Intelligence and Japan's Fifth Generation"; Edwards, *Closed World*, 298; Henrich-Franke, "EC Competition Law and the Idea of 'Open Networks' (1950s–1980s)."

69. Agar, *Science Policy under Thatcher*, 64–66.

70. "Telecom Policy" (July 19, 1982), HC Deb Vol 28 cc23–32, Hansard.

71. *Future of Telecommunications in Britain*.

72. Hills, *Deregulating Telecoms*, 129–130.

73. Agar, *Science Policy under Thatcher*, 66–67 also discusses Baker's speech in this light.

74. Kenneth Baker, "Towards an Information Economy" (September 7, 1982), T 471/45, TNA.

75. Turner, *From Counterculture to Cyberculture*; Barbrook and Cameron, "Californian Ideology."

76. Kenneth Baker, "Towards an Information Economy" (September 7, 1982), T 471/45, TNA.

77. Turner, *From Counterculture to Cyberculture*, 254.

78. Schiller, *Information and the Crisis Economy*, 45–72.

79. "Extracts from a Speech, and Answers to Questions, given by the Secretary of State to Senior BT Managers" (July 29, 1982), TCD 69/2/115, BTA.

80. George Jefferson to Patrick Jenkin, "Privatisation," April 19, 1982, TCD 69/2/45, BTA.

81. George Jefferson to Patrick Jenkin, "Privatisation," June 16, 1982, TCD 69/2/68, BTA.

82. Michael Bett, "Privatisation—Employee Shareholding" (January 1983), TCD 69/3/5, BTA.

83. "Note of the Main Points Arising at a Meeting Held at the Department of Industry" (June 15, 1983), TCD 69/3/98, BTA.

84. George Jefferson to Kenneth Baker, "BT Privatisation: Employee Share Scheme," December 12, 1983, TCD 69/3/194, BTA.

85. Newman, *Selling of British Telecom*, 150.

86. Kenneth Baker, "Towards an Information Economy" (September 7, 1982), T 471/45, TNA; Bell, *Coming of Post-Industrial Society*.

87. Webster, *Theories of the Information Society*, 38.

88. Turner, *From Counterculture to Cyberculture*, 228–230.

89. "Telecommunications Bill: 2nd Reading" (November 29, 1982), PREM 19/1100, TNA; "Screen Test."

90. "British Telecom" (November 16, 1984), HC Deb Vol 67 cc915–922, Hansard.

91. Forester, *High-Tech Society*, 92–95.

92. Lando, "European Community's Road"; European Commission, *Towards a Dynamic European Economy*.

93. European Commission, *Europe and the Global Information Society*.

94. Sweetman, *Cyber and the City*, 119.

95. Sweetman, *Cyber and the City*, 115.

96. Newman, *Selling of British Telecom*, 86–87.

97. *BT's City Business System*.

98. Jackson, "Think-Tank Archipelago"; Rollings, "Cracks in the Post-War Keynesian Settlement?"; Rollings, "Organised Business and the Rise of Neoliberalism"; Edwards, "'Manufacturing Capitalists'"; Edwards, "'Financial Consumerism.'"

99. Turner, *From Counterculture to Cyberculture*; Barbrook and Cameron, "Californian Ideology."

CONCLUSION

1. Parker, *Official History of Privatisation*, 1: The Formative Years, 1970–1987:416–417.

2. This line was inspired by Edwards, *Vast Machine*, 12.

3. Lean, *Electronic Dreams*; Lean, "Prestel."

4. Pardo-Guerra, *Automating Finance*, 124–128.

5. Lean, *Electronic Dreams*.

6. Mailland and Driscoll, *Minitel*.

7. Edgerton, *Warfare State*; Edgerton, *Rise and Fall of the British Nation*; Davies, *City of London and Social Democracy*; Ortolano, *Thatcher's Progress*.

8. Edgerton, "What Came between New Liberalism and Neoliberalism?," 37, 45.

9. "Managing Director's Committee: Telecommunications" (November 1, 1974), TCC 55/6/178, BTA.

10. McDonald, "To Corporatize or Not to Corporatize."

11. McDonald, "To Corporatize or Not to Corporatize"; Clifton and Díaz-Fuentes, "State and Public Corporations."

12. Garvey, "Artificial Intelligence and Japan's Fifth Generation"; Medina, *Cybernetic Revolutionaries*; Mailland and Driscoll, *Minitel*.

13. McDonald, "To Corporatize or Not to Corporatize"; Clifton and Díaz-Fuentes, "State and Public Corporations."

14. Frost, *Alternating Currents*.

15. Ashworth, *State in Business*, 204–205.

16. Larner and Laurie, "Travelling Technocrats, Embodied Knowledges."

17. Cook and Uchida, "Privatisation and Economic Growth"; Parker and Kirkpatrick, "Privatisation in Developing Countries."

18. Paul Ceruzzi's concept of digitalization as a "universal solvent" is presented in Haigh, "Introducing the Early Digital."

19. Hughes, "Evolution of Large Technological Systems."

BIBLIOGRAPHY

ARCHIVAL COLLECTIONS

Bank of England Archive, London, UK

British Postal Museum and Archive, Farringdon, London, UK

British Telecom Archives, Holborn, London, UK

George Washington University Special Collections, Washington, DC, US

International Telecommunications Union Archive, Geneva, Switzerland

Margaret Thatcher Foundation Archive, UK

National Air and Space Museum Archives Center, Washington DC, US

National Archives, Kew, London, UK

National Museum of American History Archives Center, Washington DC, US

PUBLISHED SOURCES

"£100m Boost for Network." *Telecom Today*, May 1988.

"A Commercial Post Office." *Financial Times*, March 1, 1967.

"A Year of Challenge and Change." *Post Office Telecommunications Journal* 19, no. 4 (1967): 10–14.

Abbate, Janet. *Inventing the Internet*. Cambridge, MA: MIT Press, 1999.

Abbate, Janet. *Recoding Gender: Women's Changing Participation in Computing*. Cambridge, MA: MIT Press, 2012.

Advisory Council on Science and Technology. *Optoelectronics: Building on Our Investment*. London: HMSO, 1988.

Agar, Jon. "Making a Meal of the Big Dish: The Construction of the Jodrell Bank Mark 1 Radio Telescope as a Stable Edifice, 1946–57." *British Journal for the History of Science* 27, no. 1 (1994): 3–21.

Agar, Jon. *Science and Spectacle: The Work of Jodrell Bank in Postwar British Culture.* Amsterdam: Routledge, 1998.

Agar, Jon. *The Government Machine: A Revolutionary History of the Computer.* Cambridge, MA: MIT Press, 2003.

Agar, Jon. *Science Policy under Thatcher.* London: UCL Press, 2019.

Agar, Jon. "What Is Science For? The Lighthill Report on Artificial Intelligence Reinterpreted." *British Journal for the History of Science* 53, no. 3 (2020): 289–310. https://doi.org/10.1017/S0007087420000230.

Agar, Jon, and Jacob Ward, eds. *Histories of Technology, the Environment and Modern Britain.* London: UCL Press, 2018. https://doi.org/10.14324/111.9781911576570.

"Age of Automation, The." *The Reith Lectures.* BBC Radio 4, 1964. https://www.bbc.co.uk/programmes/p00h9182/episodes/player.

Agic, Damir, and Nico Grove. "Role of the State in Telecommunications Infrastructure Financing across Europe: The Telephony Service from the 1880s to the First World War." In *Infrastructure Finance in Europe,* 255–281. Oxford: Oxford University Press, 2016. https://doi.org/10.1093/acprof:oso/9780198713418.003.0011.

Akera, Atsushi. "Engineers or Managers? The Systems Analysis of Electronic Data Processing in the Federal Bureaucracy." In *Systems, Experts, and Computers: The Systems Approach in Management and Engineering, World War II and After,* edited by Agatha C. Hughes and Thomas P. Hughes, 191–220. Cambridge, MA: MIT Press, 2000.

Alder, Ken. *Engineering the Revolution: Arms and Enlightenment in France, 1763–1815.* Chicago: University of Chicago Press, 1997.

Aldous, Tony. "Controlled Chaos Proves a Winner." *Chartered Surveyor Weekly* 4, no. 13 (1983): 662–663.

Aldous, Tony. "Suffolk Sensibilities." *Building Design,* no. 680 (1984): 14–15.

Alexander, Anthony. *Britain's New Towns: Garden Cities to Sustainable Communities.* London: Routledge, 2009.

"All Change at the Post Office." *Times,* August 4, 1966.

Allenstein, B. M., and D. E. Probert. "A Strategic Control Module for a Corporate Model of British Telecom." In *Forecasting Public Utilities,* edited by O. D. Anderson, 39–58. Amsterdam: North-Holland, 1980.

Andersson, Jenny. *The Future of the World: Futurology, Futurists, and the Struggle for the Post Cold War Imagination.* Oxford: Oxford University Press, 2018.

Andersson, Jenny. "The Great Future Debate and the Struggle for the World." *American Historical Review* 117, no. 5 (2012): 1411–1430.

Andersson, Jenny, and Eglė Rindzevičiūtė, eds. *The Struggle for the Long-Term in Transnational Science and Politics: Forging the Future.* London: Routledge, 2015.

"Appointment of Mr. J.H.H. Merriman as Senior Director of Engineering." *Post Office Electrical Engineers' Journal* 60, no. 2 (1967): 139.

Ardill, John. "Worker-Directors Dropped by Post Office Management." *Guardian,* December 13, 1979.

Ashworth, William. *The State in Business: 1945 to the Mid-1980s*. Basingstoke, UK: Macmillan, 1991.

Aspray, William, and Christopher Loughnane. "Foreground the Background: Business, Economics, Labor, and Government Policy as Shaping Forces in Early Digital Computing History." In *Exploring the Early Digital*, edited by Thomas Haigh, 19–40. Cham, Switzerland: Springer, 2019. https://doi.org/10.1007/978-3-030-02152-8.

"AT&T, British Telecom Set $10B Venture." *CNNMoney*, July 27, 1998. http://money .cnn.com/1998/07/27/deals/bt/.

"AT&T Sets Telstar II Launch for Spring; Lessening of Radiation Damage a Main Goal." *Wall Street Journal*, January 2, 1963.

AT&T's Universal Telephone Service. AT&T, 1989. https://www.youtube.com/watch?v =KklN1DCRmCU.

Badenoch, Alexander, and Andreas Fickers, eds. *Materializing Europe: Transnational Infrastructures and the Project of Europe*. Basingstoke, UK: Palgrave Macmillan, 2010.

Baker, Kenneth. "I. Information Technology: Industrial and Employment Opportunities." Cantor Lectures: The New Information Technology. *Journal of the Royal Society of Arts* 130, no. 5316 (November 1982): 780–790.

Baker, Kenneth. *The Turbulent Years: My Life in Politics*. London: Faber & Faber, 1993.

Balbi, Gabriele, and Andreas Fickers, eds. *History of the International Telecommunication Union: Transnational Techno-Diplomacy from the Telegraph to the Internet*. Berlin: De Gruyter Oldenbourg, 2020.

Balbi, Gabriele, and Christiane Berth. "Towards a Telephonic History of Technology." *History and Technology* 35, no. 2 (2019): 105–114. https://doi.org/10.1080 /07341512.2019.1652959.

Baran, Paul. "The Future Computer Utility." *National Affairs* 8 (Summer 1967): 75–87.

Barbrook, Richard, and Andy Cameron. "The Californian Ideology." *Science as Culture* 6, no. 1 (1996): 44–72.

Barres-Baker, M. C. "A Brief History of the London Borough of Brent." *Brent Museum and Archive Occasional Publications*, no. 5 (2007): 1–14.

Barron, D. A. "Dialling for Trunk Calls: A Cheaper Telephone Service." *Manchester Guardian*, December 4, 1958.

Barrow, Becky. "BT's Bunker Is a Brave New World in the Making." *Daily Telegraph*, March 15, 2001.

Barry, Andrew. *Political Machines: Governing a Technological Society*. London: The Athlone Press, 2001.

Barry, Andrew. "The Anti-Political Economy." In *The Technological Economy*, edited by Andrew Barry and Don Slater, 84–100. Abingdon, UK: Routledge, 2005.

Bates, O. "Ploughing the Seabed." *Post Office Telecommunications Journal* 22, no. 4 (1970): 2–4.

Bauer, Johannes M. "Changing Roles of the State in Telecommunications." *International Telecommunications Policy Review* 17, no. 1 (2010): 1–36.

Bealey, Frank. *The Post Office Engineering Union: The History of the Post Office Engineers, 1870–1970*. London: Bachman & Turner, 1976.

Beesley, M., and S. Littlechild. "Privatisation: Principles, Problems and Priorities." *Lloyds Bank Review*, no. 149 (July 1983): 1–20.

Beesley, M. E. *Privatization, Regulation and Deregulation*. London: Routledge, 1992.

Beesley, M. E., and S. C. Littlechild. "The Regulation of Privatized Monopolies in the United Kingdom." *RAND Journal of Economics* 20, no. 3 (1989): 454–472.

Beesley, Michael E. *Liberalisation of the Use of British Telecommunications Network*. London: HMSO, 1981.

Bell, Daniel. *The Coming of Post-Industrial Society: A Venture in Social Forecasting*. New York: Basic Books, 1973.

Bellringer, Christopher, and Ranald Michie. "Big Bang in the City of London: An Intentional Revolution or an Accident?" *Financial History Review* 21, no. 2 (2014): 111–137. https://doi.org/10.1017/S0968565014000092.

Beloff, Michael. *The Plateglass Universities*. London: Secker & Warburg, 1968.

Berman, Sheri. *The Primacy of Politics: Social Democracy and the Making of Europe's Twentieth Century*. Cambridge, UK: Cambridge University Press, 2006. https://doi.org /10.1017/CBO9780511791109.

Bijker, Wiebe E., Thomas P. Hughes, and Trevor Pinch, eds. *The Social Construction of Technological Systems: New Directions in the Sociology and History of Technology*. Cambridge, MA: MIT Press, 1987.

Billings, Mark, and John Wilson. "'Breaking New Ground': The National Enterprise Board, Ferranti, and Britain's Prehistory of Privatization." *Enterprise & Society* 20, no. 4 (December 2019): 907–938. https://doi.org/10.1017/eso.2019.13.

"Bill's a Winner, Says Juke-Box 'Jury.'" *BT Journal* 9, no. 2 (June 1988): 88–89.

Bishop, Matthew, and David Thompson. "Privatisation in the UK: Deregulatory Reform and Public Enterprise Performance." In *Privatisation: A Global Perspective*, edited by V. V. Ramanadham, 1–28. London: Routledge, 1993.

Blok, Aad, and Greg Downey, eds. *Uncovering Labour in Information Revolutions, 1750–2000*. International Review of Social History Supplements 11. Cambridge, UK: Cambridge University Press, 2004.

Boas, Taylor C., and Jordan Gans-Morse. "Neoliberalism: From New Liberal Philosophy to Anti-Liberal Slogan." *Studies in Comparative International Development* 44, no. 2 (2009): 137–161. https://doi.org/10.1007/s12116-009-9040-5.

Bogard, William. *The Simulation of Surveillance: Hypercontrol in Telematic Societies*. Cambridge, UK: Cambridge University Press, 1996.

Bohn, Lewis C. *Information Technology in Development*. Croton-on-Hudson, NY: Hudson Institute, 1968.

Boltanski, Luc, and Eve Chiapello. *The New Spirit of Capitalism*. Translated by Gregory Elliott. London: Verso, 2007.

Boon, Rachel. "'Research Is the Door to Tomorrow': The Post Office Engineering Research Station, 1933–1958." PhD diss., University of Manchester, 2021.

Booth, Alan. *The Management of Technical Change: Automation in the UK and USA since 1950*. Basingstoke, UK: Palgrave Macmillan, 2007.

Borup, Mads, Nik Brown, Kornelia Konrad, and Harro Van Lente. "The Sociology of Expectations in Science and Technology." *Technology Analysis & Strategic Management* 18, no. 3–4 (2006): 285–298.

Bory, Paolo, Gianluigi Negro, and Balbi Gabriele. "Introduction." In *Computer Network Histories*, edited by Paolo Bory, Gianluigi Negro, and Gabriele Balbi, 7–15. Zurich: Chronos Verlag, 2019. https://doi.org/10.33057/chronos.1539.

Bowker, Geoffrey. "How to Be Universal: Some Cybernetic Strategies, 1943–70." *Social Studies of Science* 23, no. 1 (1993): 107–127.

Bowker, Geoffrey. "Information Mythology: The World of/as Information." In *Information Acumen: The Understanding and Use of Knowledge in Modern Business*, edited by Lisa Bud-Frierman, 231–247. London: Routledge, 1994.

Bowker, Geoffrey C., and Susan Leigh Star. *Sorting Things Out: Classification and Its Consequences*. Inside Technology. Cambridge, MA: MIT Press, 1999.

Bowler, Peter J. *Science for All: The Popularization of Science in Early Twentieth-Century Britain*. Chicago: University of Chicago Press, 2009.

Bradshaw, Tancred. "The Dead Hand of the Treasury: The Economic and Social Development of the Trucial States, 1948–60." *Middle Eastern Studies* 50, no. 2 (2014): 325–342.

Braun, Wernher von. "Now at Your Service—The World's Most Talkative Satellite." *Popular Science*, May 1971, pp. 56–57, 138.

Breary, D. "A Long-Term Study of the United Kingdom Trunk Network, Part 1—General Methodology: Forecasts: Plant Study." *Post Office Electrical Engineers' Journal* 66, no. 4 (1974): 210–216.

Breary, D. "A Long-Term Study of the United Kingdom Trunk Network, Part 2—Network-Layout Studies and General Conclusions." *Post Office Electrical Engineers' Journal* 67, no. 1 (1974): 37–41.

"British Acquire 20% of MCI." *New York Times*, October 1, 1994. https://www.nytimes.com/1994/10/01/business/british-acquire-20-of-mci.html.

British Telecom International . . . It's You We Answer To. BT. YouTube video, 1988. https://www.youtube.com/watch?v=WIDrUsx1Tf8.

Brooke, Stephen. "Living in 'New Times': Historicizing 1980s Britain." *History Compass* 12, no. 1 (2014): 20–32.

Bruton, Elizabeth, and Graeme Gooday. "Listening in Combat—Surveillance Technologies beyond the Visual in the First World War." *History and Technology* 32, no. 3 (2016): 213–226.

"BT Enterprise in Venture to Exploit R&D Spin-Off." *Electronics and Power* 28, no. 2 (1982): 141.

"BT Exits Tech Mahindra." *Hindu*, December 12, 2012. http://www.thehindu.com /business/companies/bt-exits-tech-mahindra/article4191751.ece.

BT's City Business System: The Power behind the Button. BT. YouTube video, 1984. https://www.youtube.com/watch?v=1rGl6wnnWQM.

Bud, Robert. "Penicillin and the New Elizabethans." *British Journal for the History of Science* 31, no. 3 (1998): 305–333.

"Building Dossier: Martlesham Heath." *Building* 253, no. 7575 (1988): 43–54.

Butrica, Andrew J. *Beyond the Ionosphere: Fifty Years of Satellite Communication*. Washington, DC: NASA History Office, 1997.

Cable to the Continent. AT&T. YouTube video, 1959. https://www.youtube.com/watch ?v=yqRj3lvvg7Y.

Cahan, David. "The Geopolitics and Architectural Design of a Metrological Laboratory: The Physikalische-Technische Reichsanstalt in Imperial Germany." In *The Development of the Laboratory: Essays on the Place of Experiments in Industrial Civilization*, edited by Frank A. J. L. James, 137–154. Basingstoke, UK: Macmillan, 1989.

Cahill, Damien. *The End of Laissez-Faire? On the Durability of Embedded Neoliberalism*. Cheltenham, UK: Edward Elgar, 2014.

Campbell, John L., and Ove K. Pedersen, eds. *The Rise of Neoliberalism and Institutional Analysis*. Princeton, NJ: Princeton University Press, 2001.

Campbell-Kelly, Martin. *ICL: A Business and Technical History*. Oxford: Clarendon Press, 1989.

Campbell-Kelly, Martin. *From Airline Reservations to Sonic the Hedgehog: A History of the Software Industry*. Cambridge, MA: MIT Press, 2004.

Campbell-Kelly, Martin, and Daniel D. Garcia-Swartz. "Economic Perspectives on the History of the Computer Time-Sharing Industry, 1965–1985." *IEEE Annals of the History of Computing* 30, no. 1 (2008): 16–36. https://doi.org/10.1109/MAHC.2008.3.

Campbell-Kelly, Martin, and Daniel D. Garcia-Swartz. "The History of the Internet: The Missing Narratives." *Journal of Information Technology* 28, no. 1 (2013): 18–33. https://doi.org/10.1057/jit.2013.4.

Campbell-Kelly, Martin, and Daniel D. Garcia-Swartz. *From Mainframes to Smartphones: A History of the International Computer Industry*. Cambridge, MA: Harvard University Press, 2015.

Campbell-Kelly, Martin, William Aspray, Nathan Ensmenger, and Jeffrey R. Yost. *Computer: A History of the Information Machine*. 3rd ed. Boulder, CO: Westview Press, 2014.

Campbell-Smith, Duncan. *Masters of the Post: The Authorized History of the Royal Mail*. London: Penguin, 2012.

Cane, Alan. "Hewlett-Packard Purchase." *Financial Times*, May 17, 1993.

Cantor, Muriel G., and Joel M. Cantor. "Regulation and Deregulation: Telecommunication Politics in the United States." In *New Communication Technologies and the Public Interest: Comparative Perspectives on Policy and Research*, edited by Marjorie Ferguson, 84–101. London: Sage, 1986.

Capie, Forrest. *The Bank of England: 1950s to 1979*. Cambridge, UK: Cambridge University Press, 2010. https://doi.org/10.1017/CBO9780511761478.

Cassis, Youssef. Translated by Jacqueline Collier. *Capitals of Capital: The Rise and Fall of International Financial Centres 1780–2005*. Cambridge, UK: Cambridge University Press, 2006.

Cassis, Youssef, Giuseppe De Luca, and Massimo Florio, eds. *Infrastructure Finance in Europe: Insights into the History of Water, Transport, and Telecommunications*. Oxford: Oxford University Press, 2016.

Castells, Manuel. *The Rise of the Network Society*. 2nd ed. The Information Age: Economy, Society and Culture, Vol. 1. Chichester, UK: Wiley-Blackwell, 2010.

Caves, Richard, ed. *Britain's Economic Prospects*. Washington, DC: The Brookings Institution, 1968.

Cawson, Alan, Kevin Morgan, Douglas Webber, Peter Holmes, and Anne Stevens. *Hostile Brothers: Competition and Closure in the European Electronics Industry*. Oxford: Clarendon Press, 1990.

Ceruzzi, Paul E. *A History of Modern Computing*. Cambridge, MA: MIT Press, 2003.

Ceruzzi, Paul E. *Internet Alley: High Technology in Tysons Corner, 1945–2005*. Cambridge, MA: MIT Press, 2008.

Chandler, Alfred D. *The Visible Hand: Managerial Revolution in American Business*. Cambridge, MA: Belknap Press, 1977.

Childs, D. R. "Harlow." *Architects' Journal*, no. 116 (1952): 196.

Clapson, Mark. *Invincible Green Suburbs, Brave New Towns: Social Change and Urban Dispersal in Postwar England*. Manchester, UK: Manchester University Press, 1998.

Clark, Jon, Ian McLoughlin, Howard Rose, and Robin King. *The Process of Technological Change: New Technology and Social Choice in the Workplace*. Cambridge, UK: Cambridge University Press, 1988.

Clark, Pete. "Tower to the People." *Scotsman*, June 30, 1995.

Clarke, Arthur C. "Spinoff from Space." *Bell Telephone Magazine* 48, no. 5 (1969): 26–32.

Clarke, S., and A. M. Tobias. "Complexity in Corporate Modelling: A Review." *Business History* 37, no. 1 (1995): 17–44.

Clarke, Thomas. "The Political Economy of the UK Privatization Programme: A Blueprint for Other Countries?" In *The Political Economy of Privatization*, edited by Thomas Clarke and Christos Pitelis, 117–133. London: Routledge, 1993.

Clifton, Judith, and Daniel Díaz-Fuentes. "The State and Public Corporations." In *Handbook of the International Political Economy of the Corporation*, 106–119. Cheltenham, UK: Edward Elgar, 2018.

Clifton, Judith, Francisco Comín, and Daniel Díaz Fuentes. "Privatizing Public Enterprises in the European Union 1960–2002: Ideological, Pragmatic, Inevitable?"

Journal of European Public Policy 13, no. 5 (2006): 736–756. https://doi.org/10.1080 /13501760600808857.

Clifton, Judith, Pierre Lanthier, and Harm Schröter. "Regulating and Deregulating the Public Utilities 1830–2010." *Business History* 53, no. 5 (2011): 659–672.

Coase, R. H. "The Lighthouse in Economics." *Journal of Law and Economics* 17, no. 2 (1974): 357–376. https://doi.org/10.1086/466796.

Cochrane, Peter. "Home / Personal / About / CV." Peter Cochrane. Accessed November 3, 2022. https://petercochrane.com/personal/about.

Cole, A. C. "The Computer Exchange." *Post Office Telecommunications Journal* 22, no. 1 (1970): 22–23.

Cole, Tim. "About Britain: Driving the Landscape of Britain (at Speed?)." In *Histories of Technology, the Environment and Modern Britain*, 123–141. London: UCL Press, 2018.

Collins, Martin. *A Telephone for the World: Iridium, Motorola, and the Making of a Global Age*. Baltimore, MD: Johns Hopkins University Press, 2018.

Collins, Martin. "One World . . . One Telephone: Iridium, One Look at the Making of a Global Age." *History and Technology* 21, no. 3 (2005): 301–324. https://doi.org /10.1080/07341510500205449.

Committee of Inquiry on the Future of Broadcasting, and Noel Annan. *Report of the Committee on the Future of Broadcasting*. Cmnd. 6753. London: HMSO, 1977.

Committee on Privacy. *Report of the Committee on Privacy*. Cmnd. 5012. London: HMSO, 1972.

"Communications Explosion: Early Bird—and After, The." *TIME*, May 14, 1965.

Communications Steering Group. *The Infrastructure for Tomorrow*. London: HMSO, 1988.

"Companies | Innovation Martlesham." Accessed July 21, 2017. http://www.innova tionmartlesham.com/companies/.

"Computer Centre in Kensington, The." *Post Office Telecommunications Journal* 17, no. 1 (1965): 6–8.

Cook, Paul, and Yuichiro Uchida. "Privatisation and Economic Growth in Developing Countries." *Journal of Development Studies* 39, no. 6 (2003): 121–154. https://doi .org/10.1080/00220380312331293607.

Cookson, Clive. "BT to Computerize Directory Inquiries." *The Times*, May 13, 1983.

Coopey, Richard. "Industrial Policy in the White Heat of the Scientific Revolution." In *The Wilson Governments, 1964–1970*, edited by Nick Tiratsoo, Richard Coopey, and Steven Fielding, 102–122. London: Pinter, 1993.

Copeland, B. Jack. "Colossus and the Rise of the Modern Computer." In *Colossus: The Secrets of Bletchley Park's Codebreaking Computers*, edited by B. Jack Copeland, 101–115. Oxford: Oxford University Press, 2006.

Copeland, B. Jack, ed. *Colossus: The Secrets of Bletchley Park's Codebreaking Computers*. Oxford: Oxford University Press, 2006.

Copeland, B. Jack. "Machine against Machine." In *Colossus: The Secrets of Bletchley Park's Codebreaking Computers*, edited by B. Jack Copeland, 64–77. Oxford: Oxford University Press, 2006.

"Corning in Deal with NetOptix, 2 Other Firms." *Los Angeles Times*, February 15, 2000. http://articles.latimes.com/2000/feb/15/business/fi-64359.

Cortada, James W. *The Digital Hand: How Computers Changed the Work of American Manufacturing, Transportation, and Retail Industries*. Oxford: Oxford University Press, 2004.

Cortada, James W. *IBM: The Rise and Fall and Reinvention of a Global Icon*. Cambridge, MA: MIT Press, 2019.

Cosier, J. E. H. "Getting to Grips with Undersea Cables." *Post Office Telecommunications Journal* 30, no. 1 (1978): 7–8.

Craig, Cleo F. "Equipping Ourselves for Today's Responsibilities." *Bell Telephone Magazine* 34, no. 4 (1955): 217–224.

Crawford, A. B., C. C. Cutler, R. Kompfner, and L. C. Tillotson. "The Research Background of the Telstar Experiment." *Bell System Technical Journal* 42, no. 4 (1963): 747–751.

Crawford, Susan. *Captive Audience: The Telecom Industry and Monopoly Power in the New Gilded Age*. New Haven, CT: Yale University Press, 2014.

"'Crazed Communist Scheme,' PM Johnson Says of Corbyn's Plan for BT." *Reuters*, November 15, 2019. https://www.reuters.com/article/us-britain-election-bt-johnson -idUSKBN1XP1ER.

Cronin, James. *The Politics of State Expansion: War, State and Society in Twentieth Century Britain*. London: Routledge, 1991.

Dalibert, Marion, and Simona De Iulio. "The Representational Intertwinement of Gender, Age and Uses of Information and Communication Technology: A Comparison between German and French Preteen Magazines." In *Connecting Women: Women, Gender and ICT in Europe in the Nineteenth and Twentieth Century*, edited by Valérie Schafer and Benjamin G. Thierry, 89–100. Cham, Switzerland: Springer, 2015.

"Dan Dare, Pilot of the Future, in The Big City Caper." *Eagle and Swift*, May 23, 1964.

Davids, Mila. "The Privatisation and Liberalisation of Dutch Telecommunications in the 1980s." *Business History* 47, no. 2 (2005): 219–243. https://doi.org/10.1080 /0007679042000313666.

Davies, Aled. *The City of London and Social Democracy: The Political Economy of Finance in Britain, 1959—1979*. Oxford: Oxford University Press, 2017.

Davies, Aled. "The Roots of Britain's Financialised Political Economy." In *The Neoliberal Age? Britain since the 1970s*, edited by Aled Davies, Ben Jackson, and Florence Sutcliffe-Braithwaite, 299–318. London: UCL Press, 2021. https://doi.org/10.14324 /111.9781787356856.

DeNardis, Laura. *Protocol Politics: The Globalization of Internet Governance*. Cambridge, MA: MIT Press, 2009.

Dennis, Michael Aaron. "Accounting for Research: New Histories of Corporate Laboratories and the Social History of American Science." *Social Studies of Science* 17, no. 3 (1987): 479–518.

Department of Industry. *The Post Office.* Cmnd. 7292. London: HMSO, 1978.

Department of Trade and Industry. *Competition and Choice: Telecommunications Policy for the 1990s.* Cm. 1461. London: HMSO, 1991.

Dewandre, Nicole. "Europe and New Communication Technologies." In *New Communication Technologies and the Public Interest: Comparative Perspectives on Policy and Research,* edited by Marjorie Ferguson, 137–149. London: Sage, 1986.

Doubleday, C. F., and D. E. Probert. "The Development of an Integrated Communications Demand Model for the British Telecommunications Business." *Journal of the Operational Research Society* 36, no. 12 (1985): 1083–1093.

Downey, Greg. "Virtual Webs, Physical Technologies, and Hidden Workers: The Spaces of Labor in Information Internetworks." *Technology and Culture* 42, no. 2 (2001): 209–235.

Downey, Gregory J. *Telegraph Messenger Boys: Labor, Technology, and Geography, 1850–1950.* London: Routledge, 2002.

Duménil, Gérard, and Dominique Lévy. Translated by Derek Jeffers. *Capital Resurgent: Roots of the Neoliberal Revolution.* Cambridge, MA: Harvard University Press, 2004.

Duncan, C. C. "Communications and Defense." *Bell Telephone Magazine* 37, no. 1 (1958): 15–24.

Dyson, Esther, George Gilder, George Keyworth, and Alvin Toffler. "Cyberspace and the American Dream: A Magna Carta for the Knowledge Age (August 22, 1994, Future Insight Release 1.2, Progress & Freedom Foundation)." *Information Society* 12, no. 3 (1996): 295–308.

Edge, David. "Technological Metaphor and Social Control." *New Literary History* 6, no. 1 (1974): 135–147. https://doi.org/10.2307/468345.

Edgerton, David. *Science, Technology and the British Industrial "Decline," 1870–1970.* Cambridge, UK: Cambridge University Press, 1996.

Edgerton, David. "The 'White Heat' Revisited: The British Government and Technology in the 1960s." *Twentieth Century British History* 7, no. 1 (1996): 53–82.

Edgerton, David. *Warfare State: Britain, 1920–1970.* Cambridge, UK: Cambridge University Press, 2005.

Edgerton, David. *Shock of the Old: Technology and Global History since 1900.* London: Profile Books, 2006.

Edgerton, David. "The Contradictions of Techno-Nationalism and Techno-Globalism: A Historical Perspective." *New Global Studies* 1, no. 1 (2007). https://doi.org/10.2202/1940-0004.1013.

Edgerton, David. *England and the Aeroplane: Militarism, Modernity and Machines.* London: Penguin, 2013.

Edgerton, David. *The Rise and Fall of the British Nation: A Twentieth-Century History*. London: Penguin, 2019.

Edgerton, David. "What Came between New Liberalism and Neoliberalism? Rethinking Keynesianism, the Welfare State and Social Democracy." In *The Neoliberal Age? Britain since the 1970s*, edited by Aled Davies, Ben Jackson, and Florence Sutcliffe-Braithwaite, 30–51. London: UCL Press, 2021. https://doi.org/10.14324/111.9781787356856.

Edwards, Amy. "'Manufacturing Capitalists': The Wider Share Ownership Council and the Problem of 'Popular Capitalism,' 1958–92." *Twentieth Century British History* 27, no. 1 (2016): 100–123.

Edwards, Amy. "'Financial Consumerism': Citizenship, Consumerism and Capital Ownership in the 1980s." *Contemporary British History* 31, no. 2 (2017): 210–229.

Edwards, Paul N. *The Closed World: Computers and the Politics of Discourse in Cold War America*. Cambridge, MA: MIT Press, 1996.

Edwards, Paul N. "Infrastructure and Modernity: Force, Time, and Social Organisation in the History of Sociotechnical Systems." In *Modernity and Technology*, edited by Thomas J. Misa, Philip Brey, and Andrew Feenberg, 185–224. Cambridge, MA: MIT Press, 2003.

Edwards, Paul N. *A Vast Machine: Computer Models, Climate Data, and the Politics of Global Warming*. Cambridge, MA: MIT Press, 2010.

Edwards, Paul N., and Gabrielle Hecht. "History and the Technopolitics of Identity: The Case of Apartheid South Africa." *Journal of Southern African Studies* 36, no. 3 (2010): 619–639. https://doi.org/10.1080/03057070.2010.507568.

Egan, Michael. "A Separate Company for a Broadband Network." *The Age*, September 5, 2008, sec. Business. https://www.theage.com.au/business/a-separate-company-for-a-broadband-network-20080904-49zt.html.

Elam, Mark. "National Imaginations and Systems of Innovation." In *Systems of Innovation: Technologies, Institutions, and Organizations*, edited by Charles Edquist, 157–173. Science, Technology and the International Political Economy Series. London: Pinter, 1997.

Elmendorf, C. H., and B. C. Heezen. "Oceanographic Information for Engineering Submarine Cable Systems." *Bell System Technical Journal* 36, no. 5 (1957): 1035–1094.

Ensmenger, Nathan. "'Beards, Sandals, and Other Signs of Rugged Individualism': Masculine Culture within the Computing Professions." *Osiris* 30, no. 1 (2015): 38–65.

Ensmenger, Nathan. *The Computer Boys Take Over: Computers, Programmers, and the Politics of Technical Expertise*. Cambridge, MA: MIT Press, 2010.

Ensmenger, Nathan. "The Digital Construction of Technology: Rethinking the History of Computers in Society." *Technology and Culture* 53, no. 4 (2012): 753–776. https://doi.org/10.1353/tech.2012.0126.

Esteves, Olivier. *The "Desegregation" of English Schools: Bussing, Race and Urban Space, 1960s–1980s*. Manchester, UK: Manchester University Press, 2018.

European Commission. *Europe and the Global Information Society: Recommendations to the European Council.* Brussels: EU Publications Office, 1995.

European Commission. *Towards a Dynamic European Economy: Green Paper on the Development of the Common Market for Telecommunications Services and Equipment.* COM(87) 290. Brussels: European Commission, 1987.

"Facts from Space via Telstar." *Bell Telephone Magazine* 41, no. 3 (1962): 38–39.

Fagan, Mary. "Disconnecting Staff to Reach the Right Number." *The Independent,* August 2, 1994.

Fari, Simone. "Telegraphic Diplomacy from the Origins to the Formative Years of the ITU, 1849–1875." In *History of the International Telecommunication Union: Transnational Techno-Diplomacy from the Telegraph to the Internet,* edited by Gabriele Balbi and Andreas Fickers, 169–190. Berlin: De Gruyter Oldenbourg, 2020.

Feigenbaum, James, and Daniel P. Gross. "Answering the Call of Automation: How the Labor Market Adjusted to the Mechanization of Telephone Operation." Working Paper Series. National Bureau of Economic Research, November 2020. https://doi.org/10.3386/w28061.

Feigenbaum, James J., and Daniel P. Gross. "Organizational Frictions and Increasing Returns to Automation: Lessons from AT&T in the Twentieth Century." Working Paper Series. National Bureau of Economic Research, December 2021. https://doi.org/10.3386/w29580.

Fensom, Harry. "How Colossus Was Built and Operated—One of Its Engineers Reveals Its Secrets." In *Colossus: The Secrets of Bletchley Park's Codebreaking Computers,* edited by B. Jack Copeland, 91–100. Oxford: Oxford University Press, 2006.

Fickers, Andreas, and Pascal Griset. *Communicating Europe: Technologies, Information, Events.* London: Palgrave Macmillan, 2019.

Fickers, Andreas, and Suzanne Lommers. "Eventing Europe: Broadcasting and the Mediated Performances of Europe." In *Materializing Europe: Transnational Infrastructures and the Project of Europe,* edited by Alexander Badenoch and Andreas Fickers, 225–251. Houndmills, Basingstoke: Palgrave Macmillan, 2010.

"Fight for the New Telephone System, The." *Financial Times,* January 28, 1972.

Finn, Bernard, and Daqing Yang, eds. *Communications Under the Seas: The Evolving Cable Network and Its Implications.* Cambridge, MA: MIT Press, 2009.

"First Intelsat 4 Placed in Orbit." *Aviation Week,* February 1, 1971.

First Report from the Select Committee on Nationalised Industries: The Post Office. Vol. 1, *Report and Proceedings of the Committee.* London: HMSO, 1967.

Fischer, Claude S. "'Touch Someone': The Telephone Industry Discovers Sociability." *Technology and Culture* 29, no. 1 (1988): 32.

Fischer, Claude S. *America Calling: A Social History of the Telephone to 1940.* Berkeley: University of California Press, 1994.

Fjaestad, Maja. "Fast Breeder Reactors in Sweden: Vision and Reality." *Technology and Culture* 56, no. 1 (2015): 86–114.

Fletcher, Richard, and Mary Fagan. "BT Sets up £2bn Venture Arm." *Telegraph*, November 19, 2000. http://www.telegraph.co.uk/finance/4472832/BT-sets-up-2bn -venture-arm.html.

Flowers, Thomas H. "D-Day at Bletchley Park." In *Colossus: The Secrets of Bletchley Park's Codebreaking Computers*, edited by B. Jack Copeland, 78–83. Oxford: Oxford University Press, 2006.

Floyd, C. F. "The Design of Martlesham Research Centre: Part 1—Basic Design Requirements and Design of Buildings." *Post Office Electrical Engineers' Journal* 69, no. 3 (1976): 146–153.

Floyd, C. F. "The Design of Martlesham Research Centre: Part 2—Services Provided." *Post Office Electrical Engineers' Journal* 69, no. 4 (1977): 258–264.

Foreman-Peck, James, and Robert Millward. *Public and Private Ownership of British Industry, 1820–1990*. Oxford: Clarendon Press, 1994.

Forester, Tom. *High-Tech Society: The Story of the Information Technology Revolution*. Cambridge, MA: MIT Press, 1987.

Forrester, Jay Wright. "Industrial Dynamics." *Harvard Business Review* 36, no. 4 (1958): 37–66.

Foster, Christopher. "Michael Beesley." *Guardian*, October 8, 1999.

Fourcade-Gourinchas, Marion, and Sarah L. Babb. "The Rebirth of the Liberal Creed: Paths to Neoliberalism in Four Countries." *American Journal of Sociology* 108, no. 3 (2002): 533–579.

"Freedom for the GPO." *Guardian*, August 4, 1966.

Frost, Robert L. *Alternating Currents: Nationalized Power in France, 1946–70*. Ithaca, NY: Cornell University Press, 1991.

"Full Speed Ahead for Cable TV—Brittan." *Guardian*, July 1, 1983.

Future of Telecommunications in Britain, The. Cmnd. 8610. London: HMSO, 1982.

Gabor, Dennis. *Inventing the Future*. London: Secker & Warburg, 1963.

Galambos, Louis. "Technology, Political Economy, and Professionalization: Central Themes of the Organizational Synthesis." *Business History Review* 57, no. 4 (1983): 471–493.

Garvey, Colin. "Artificial Intelligence and Japan's Fifth Generation: The Information Society, Neoliberalism, and Alternative Modernities." *Pacific Historical Review* 88, no. 4 (2019): 619–658. https://doi.org/10.1525/phr.2019.88.4.619.

Giddens, Anthony. *Social Theory and Modern Sociology*. Cambridge, UK: Polity Press, 1987.

Gillies, James, and Robert Cailliau. *How the Web Was Born: The Story of the World Wide Web*. Oxford: Oxford University Press, 2000.

"Girl with the Golden Voice, The." *Post Office Magazine*, August 1935.

Gleichmann, T. F., A. H. Lince, M. C. Wooley, and F. J. Braga. "Repeater Design for the North Atlantic Link." *Bell System Technical Journal* 36, no. 1 (1957): 69–101.

"Go-Ahead for TAT-8." *BT Journal* 4 (Winter 1984): 35.

Godin, Benoît. "Research and Development: How the 'D' Got into R&D." *Science and Public Policy* 33, no. 1 (2006): 59–76.

Godin, Benoît, and Désirée Schauz. "The Changing Identity of Research: A Cultural and Conceptual History." *History of Science* 54, no. 3 (2016): 276–306.

Godsmark, Chris. "BT Plans Radical Internal Shake-Up." *Independent*, October 23, 1996, sec. News. https://www.independent.co.uk/news/business/bt-plans-radical -internal-shakeup-1359907.html.

Goldie, C. T. "'Radio Campanile': Sixties Modernity, the Post Office Tower and Public Space." *Journal of Design History* 24, no. 3 (2011): 207–222.

Goldsmith, Jack, and Tim Wu. *Who Controls the Internet?: Illusions of a Borderless World*. New York: Oxford University Press, 2008.

Gooday, Graeme. "Re-writing the 'Book of Blots': Critical Reflections on Histories of Technological 'Failure.'" *History and Technology* 14, no. 4 (1998): 265–291.

Gooday, Graeme, and James Sumner, eds. *By Whose Standards? Standardization, Stability and Uniformity in the History of Information and Electrical Technologies*. Vol. 28, *History of Technology*. London: Continuum, 2008.

Gore, Jr., Al. "Remarks at Superhighway Summit." Presented at the Superhighway Summit, Royce Hall, UCLA, Los Angeles, January 11, 1994. https://clinton1.nara.gov /White_House/EOP/OVP/other/superhig.html.

"GRACE" in Action. British Movietone, 1958. YouTube. https://www.youtube.com /watch?v=YUy1f1eGNew.

Green, E. H. H. "Thatcherism: An Historical Perspective." *Transactions of the Royal Historical Society* 9 (1999): 17. https://doi.org/10.2307/3679391.

Green, E. H. H. *Ideologies of Conservatism: Conservative Political Ideas in the Twentieth Century*. Oxford: Oxford University Press, 2002.

Green, Venus. "Goodbye Central: Automation and the Decline of 'Personal Service' in the Bell System, 1878–1921." *Technology and Culture* 36, no. 4 (1995): 912.

Green, Venus. "Race, Gender, and National Identity in the American and British Telephone Industries." *International Review of Social History* 46, no. 2 (2001): 185–205.

Green, Venus. *Race on the Line: Gender, Race, & Technology in the Bell System, 1880–1980*. Durham, NC: Duke University Press, 2001.

Gribben, Roland. "50,000 at BT Interested in Redundancy." *Daily Telegraph*, May 30, 1992.

Griset, Pascal. "Innovation and Radio Industry in Europe during the Interwar Period." In *Innovations in the European Economy between the Wars*, edited by Francois Caron, Paul Erker, and Wolfram Fischer, 37–66. Berlin: De Gruyter, 1995. https:// www.degruyter.com/document/doi/10.1515/9783110881417.37/html.

Guldi, Jo. *Roads to Power: Britain Invents the Infrastructure State*. Cambridge, MA: Harvard University Press, 2012.

Habara, Kohei. "ISDN: A Look at the Future through the Past." *IEEE Communications Magazine* 26, no. 11 (1988): 25–32. https://doi.org/10.1109/35.9126.

Haigh, Alice Elena. "'To Strive, to Seek, to Find': The Origins and Establishment of the British Post Office Engineering Research Station at Dollis Hill, 1908–1938." PhD diss., University of Leeds, 2020.

Haigh, Thomas. "Computing the American Way: Contextualizing the Early US Computer Industry." *IEEE Annals of the History of Computing* 32, no. 2 (2010): 8–20. https://doi.org/10.1109/MAHC.2010.33.

Haigh, Thomas. "Inventing Information Systems: The Systems Men and the Computer, 1950–1968." *Business History Review* 75, no. 1 (2001): 15–61.

Haigh, Thomas. "The History of Information Technology." *Annual Review of Information Science and Technology* 45, no. 1 (2011): 431–487.

Haigh, Thomas. "Introducing the Early Digital." In *Exploring the Early Digital*, edited by Thomas Haigh, 1–18. Cham, Switzerland: Springer, 2019. https://doi.org/10.1007/978-3-030-02152-8.

Haigh, Thomas, and Mark Priestley. "Colossus and Programmability." *IEEE Annals of the History of Computing* 40, no. 4 (2018): 5–27.

Hale, Constance, ed. *Wired Style: Principles of English Usage in the Digital Age*. San Francisco: Wired Books, 1997.

Hall, Peter. *Cities of Tomorrow: An Intellectual History of Urban Planning and Design Since 1880*. 4th ed. Chichester, UK: Wiley-Blackwell, 2014.

Hamblin, Jacob Darwin. *Arming Mother Nature: The Birth of Catastrophic Environmentalism*. Oxford: Oxford University Press, 2013.

Hamilton, Adrian. "Joseph to Split Post Office." *Observer*, September 9, 1979.

Harper, J. M. *Monopoly and Competition in British Telecommunications: The Past, the Present and the Future*. London: Pinter, 1997.

Harris, L. Roy F. *Automatic Switching in the UK*. Sunbury, UK: The Communications Network, 2005.

Harvey, David. *The Condition of Postmodernity: An Enquiry into the Origins of Cultural Change*. Oxford: Blackwell, 1990.

Hay, Colin. "Whatever Happened to Thatcherism?" *Political Studies Review* 5, no. 2 (2007): 183–201. https://doi.org/10.1111/j.1478-9299.2007.00128.x.

Head, R. V. "Getting Sabre off the Ground." *IEEE Annals of the History of Computing* 24, no. 4 (2002): 32–39. https://doi.org/10.1109/MAHC.2002.1114868.

Headrick, Daniel R. *The Invisible Weapon: Telecommunications and International Politics, 1851–1945*. Oxford: Oxford University Press, 1991.

Hecht, Gabrielle. *The Radiance of France: Nuclear Power and National Identity after World War II*. Cambridge, MA: MIT Press, 1998.

Hecht, Gabrielle. "Planning a Technological Nation: Systems Thinking and the Politics of National Identity in Postwar France." In *Systems, Experts, and Computers: The*

Systems Approach in Management and Engineering, World War II and After, edited by Agatha C. Hughes and Thomas Parke Hughes, 133–160. Cambridge, MA: MIT Press, 2000.

Hecht, Jeff. *City of Light: The Story of Fiber Optics*. Oxford: Oxford University Press, 1999.

Henderson, P. D. "Two British Errors: Their Probable Size and Some Possible Lessons." *Oxford Economic Papers* 29, no. 2 (1977): 159–205.

Hendry, John. *Innovating for Failure: Government Policy and the Early British Computer Industry*. Cambridge, MA: MIT Press, 1990.

Hennessy, Peter. *Whitehall*. London: Secker & Warburg, 1989.

Hennessy, Peter, Susan Morrison, and Richard Townsend. "Routine Punctuated by Orgies: The Central Policy Review Staff, 1970–83." *Strathclyde Papers on Government and Politics*, no. 31 (1985).

Henrich-Franke, Christian. "Comparing Cultures of Expert Regulation: Governing Cross-Border Infrastructures." *Contemporary European History* 27, no. 2 (2018): 280–300. https://doi.org/10.1017/S0960777318000139.

Henrich-Franke, Christian. "Computer Networks on Copper Cables." In *Computer Network Histories*, edited by Paolo Bory, Gianluigi Negro, and Gabriele Balbi, 65–79. Zurich: Chronos Verlag, 2019. https://doi.org/10.33057/chronos.1539.

Henrich-Franke, Christian. "EC Competition Law and the Idea of 'Open Networks' (1950s–1980s)." *Internet Histories* 4, no. 2 (2020): 125–141. https://doi.org/10.1080/24701475.2020.1743045.

Henrich-Franke, Christian, and Léonard Laborie. "European Union for and by Communication Networks: Continuities and Discontinuities during the Second World War." *Comparativ* 28, no. 1 (2018): 82–100. https://doi.org/10.26014/j.comp.2018.01.05.

Hicks, Mar. *Programmed Inequality: How Britain Discarded Women Technologists and Lost Its Edge in Computing*. Cambridge, MA: MIT Press, 2017.

Higuchi, Toshihiro. "Atmospheric Nuclear Weapons Testing and the Debate on Risk Knowledge in Cold War America, 1945–1963." In *Environmental Histories of the Cold War*, edited by J. R. McNeill and Corinna R. Unger, 301–322. Cambridge, UK: Cambridge University Press, 2010.

Hills, Jill. *Deregulating Telecoms: Competition and Control in the United States, Japan, and Britain*. London: Praeger, 1986.

Hills, Jill. "Regulation, Innovation and Market Structure in International Telecommunications: The Case of the 1956 TAT1 Submarine Cable." *Business History* 49, no. 6 (2007): 868–885. https://doi.org/10.1080/00076790701710373.

Hills, Jill. *Telecommunications and Empire*. Urbana: University of Illinois Press, 2008.

Hindmarch-Watson, Katie. *Serving a Wired World: London's Telecommunications Workers and the Making of an Information Capital*. Oakland: University of California Press, 2020.

Högselius, Per, Arne Kaijser, and Erik van der Vleuten. *Europe's Infrastructure Transition: Economy, War, Nature.* Basingstoke, UK: Palgrave Macmillan, 2015.

Home Office. *Report of the Inquiry into Cable Expansion and Broadcasting Policy.* Cmnd. 8679. London: HMSO, 1982.

Home Office. *The Development of Cable Systems and Services.* Cmnd. 8866. London: HMSO, 1983.

Hu, Yong, and Benjamin Peters. "A Conversation on Network Histories." In *Computer Network Histories*, edited by Paolo Bory, Gianluigi Negro, and Gabriele Balbi, 115–123. Zurich: Chronos Verlag, 2019. https://doi.org/10.33057/chronos.1539.

"Huawei: UK Bans New 5G Network Equipment from September." *Guardian*, November 30, 2020. https://www.theguardian.com/technology/2020/nov/30/huawei-uk-bans-new-5g-network-equipment-from-september.

Hughes, Thomas P. "The Electrification of America: The System Builders." *Technology and Culture* 20, no. 1 (1979): 124–161. https://doi.org/10.2307/3103115.

Hughes, Thomas P. *Networks of Power: Electrification in Western Society, 1880–1930.* Baltimore, MD: Johns Hopkins University Press, 1983.

Hughes, Thomas P. "The Evolution of Large Technological Systems." In *The Social Construction of Technological Systems: New Directions in the Sociology and History of Technology*, edited by Wiebe E. Bijker, Thomas P. Hughes, and Trevor Pinch, 51–82. Cambridge, MA: MIT Press, 1987.

Hulsink, Willem. *Privatisation and Liberalisation in European Telecommunications: Comparing Britain, the Netherlands and France.* London: Routledge, 1998.

Husband, John, and Terry Pattinson. "BT Is Pulling the Plug on 6,500 Jobs." *Daily Mirror*, March 5, 1991.

Hutton, Will. "Why Britain Is So Slow in Getting a Fibre Optic Network." *Guardian*, July 28, 1994.

Huurdeman, Anton A. *The Worldwide History of Telecommunications.* Hoboken, NJ: John Wiley & Sons, 2003.

"Industrial Democracy 'Test Case.'" *Guardian*, May 17, 1977.

Information Technology Advisory Panel. *Report on Cable Systems.* London: HMSO, 1982.

International Telegraph and Telephone Consultative Committee, The. *Fifth Plenary Assembly: Green Book.* Vol. 3, *Line Transmission.* Geneva: International Telecommunication Union, 1973.

"Introducing You to LEAPS." *Post Office Magazine*, October 1957, p. 300.

"It's Lonely Up There at the Top." *Independent*, November 16, 1994.

Jack, J. S., C. W. H. Leech, and H. A. Lewis. "Route Selection and Cable Laying for the Transatlantic Cable System." *Bell System Technical Journal* 36, no. 1 (1957): 293–326.

Jackson, Ben. "The Think-Tank Archipelago: Thatcherism and Neo-Liberalism." In *Making Thatcher's Britain*, edited by Ben Jackson and Robert Saunders, 43–61. Cambridge, UK: Cambridge University Press, 2012.

Jencks, Charles. *The New Paradigm in Architecture: The Language of Post-Modernism.* New Haven, CT: Yale University Press, 2002.

John, Richard R. *Spreading the News: The American Postal System from Franklin to Morse.* Cambridge, MA: Harvard University Press, 1995.

John, Richard R. *Network Nation.* Cambridge, MA: Harvard University Press, 2010.

John, Richard R., and Léonard Laborie. "'Circuits of Victory': How the First World War Shaped the Political Economy of the Telephone in the United States and France." *History and Technology* 35, no. 2 (2019): 115–137. https://doi.org/10.7916/d8-nqx2-bm33.

Jones, D. G., and P. J. Edwards. "The Post Office Network of Radio-Relay Stations. Part 1—Radio-Relay Links and Network Planning." *Post Office Electrical Engineers' Journal* 57, no. 3 (1964): 147–155.

Jones, D. G., and P. J. Edwards. "Post Office Tower, London, and the United Kingdom Network of Microwave Links." *Post Office Electrical Engineers' Journal* 58, no. 3 (1965): 149–159.

Joyce, Patrick. *The State of Freedom: A Social History of the British State since 1800.* Cambridge, UK: Cambridge University Press, 2013.

Kahn, Herman, and Anthony J. Wiener. *The Year 2000: A Framework for Speculation on the Next Thirty-Three Years.* New York: Macmillan, 1967.

Kaijser, Arne. "The Helping Hand: In Search of a Swedish Institutional Regime for Infrastructural Systems." In *Institutions in the Transport and Communications Industries: State and Private Actors in the Making of Institutional Patterns, 1850–1990*, edited by Lena Andersson-Skog and Olle Krantz, 223–244. Nantucket, MA: Science History Publications, 1999.

Karas, Serkan, and Stathis Arapostathis. "Harbours of Crisis and Consent: The Technopolitics of Coastal Infrastructure in Colonial Cyprus, 1895–1908." *Journal of Transport History* 37, no. 2 (2016): 214–235. https://doi.org/10.1177/0022526616667365.

Kargon, Robert H., and Arthur P. Molella. *Invented Edens: Techno-Cities of the Twentieth Century.* Cambridge, MA: MIT Press, 2008.

Keegan, Victor. "Through a Glass Fibre Darkly." *Guardian*, May 22, 1993.

Kelly, M. J., S. G. Radley, G. W. Gilman, and R. J. Halsey. "A Transatlantic Telephone Cable." *Transactions of the American Institute of Electrical Engineers, Part I: Communication and Electronics* 74, no. 1 (1955): 124–139.

Kelsey, Tom. "Picking Losers: Concorde, Nuclear Power, and Their Opponents in Post-War Britain, 1954–1995." PhD diss., King's College London, 2020.

Kennedy, Devin. "The Machine in the Market: Computers and the Infrastructure of Price at the New York Stock Exchange, 1965–1975." *Social Studies of Science* 47, no. 6 (2017): 888–917.

Kennedy, Paul M. "Imperial Cable Communications and Strategy, 1870–1914." *English Historical Review* 86, no. 341 (1971): 728–752.

Kerswell, B. R., and W. G. T. Jones. "Conclusions from the Empress Digital Tandem Exchange Field Trial." *Post Office Electrical Engineers' Journal* 72, no. 1 (1979): 9–14.

Kipling, Rudyard. "The Deep Sea Cables." In *The Seven Seas*. London: Methuen, 1896.

Kline, Ronald. "Inventing an Analog Past and a Digital Future in Computing." In *Exploring the Early Digital*, edited by Thomas Haigh, 19–40. Cham, Switzerland: Springer, 2019. https://doi.org/10.1007/978-3-030-02152-8.

Kline, Ronald, and Trevor Pinch. "Users as Agents of Technological Change: The Social Construction of the Automobile in the Rural United States." *Technology and Culture* 37, no. 4 (1996): 763–795. https://doi.org/10.2307/3107097.

Kline, Ronald R. *The Cybernetics Moment: Or Why We Call Our Age the Information Age*. Baltimore, MD: Johns Hopkins University Press, 2015.

Knorr-Cetina, Karin. "From Pipes to Scopes: The Flow Architecture of Financial Markets." In *The Technological Economy*, edited by Don Slater and Andrew Barry, 122–141. London: Routledge, 2005.

Knowles, Scott G., and Stuart W. Leslie. "'Industrial Versailles': Eero Saarinen's Corporate Campuses for GM, IBM, and AT&T." *Isis* 92, no. 1 (2001): 1–33.

Kohlrausch, Martin, and Helmuth Trischler. *Building Europe on Expertise: Innovators, Organizers, Networkers*. Making Europe: Technology and Transformations. Basingstoke, UK: Palgrave Macmillan, 2014.

Krige, John. *American Hegemony and the Postwar Reconstruction of Science in Europe*. Cambridge, MA: MIT Press, 2006.

Kumar, Krishan. *Utopia and Anti-Utopia in Modern Times*. Oxford: Basil Blackwell, 1987.

Kynaston, David. *The City of London*. Vol. 4, *Club No More, 1945–2000*. London: Pimlico, 2002.

Laborie, Léonard. "Fragile Links, Frozen Identities: The Governance of Telecommunication Networks and Europe (1944–53)." *History and Technology* 27, no. 3 (2011): 311–330. https://doi.org/10.1080/07341512.2011.604175.

Lamberton, Donald, ed. "The Information Revolution." *Annals of the American Academy of Political and Social Science*, Vol. 412, 1974. https://www.jstor.org/stable/i24302.

Lando, Steven Dov. "The European Community's Road to Telecommunications Deregulation." *Fordham Law Review* 62, no. 7 (1994): 2159–2198.

Larger, Peter. "BT Halts Move to Optical Cables." *Guardian*, February 17, 1990.

Larner, Wendy, and Nina Laurie. "Travelling Technocrats, Embodied Knowledges: Globalising Privatisation in Telecoms and Water." *Geoforum* 41, no. 2 (2010): 218–226.

Launius, Roger D. "Writing the History of Space's Extreme Environment." *Environmental History* 15, no. 3 (2010): 526–532.

Law, John. "Technology and Heterogeneous Engineering: The Case of Portuguese Expansion." In *The Social Construction of Technological Systems: New Directions in the Sociology and History of Technology*, edited by Wiebe E. Bijker, Thomas P. Hughes, and Trevor Pinch, 111–134. Cambridge, MA: MIT Press, 1987.

Lawson, Nigel. *The View from No. 11: Memoirs of a Tory Radical*. London: Bantam Press, 1992.

Lean, Tom. *Electronic Dreams: How 1980s Britain Learned to Love the Computer.* London: Bloomsbury Sigma, 2016.

Lean, Tom. "Prestel: The British Internet That Never Was." *History Today*, August 23, 2016. https://www.historytoday.com/history-matters/prestel-british-internet-never -was.

Lécuyer, Christophe. *Making Silicon Valley: Innovation and the Growth of High Tech, 1930–1970.* Cambridge, MA: MIT Press, 2005.

Ledger, Robert. "'A Transition from Here to There?': Neo-Liberal Thought and Thatcherism." PhD diss., Queen Mary, University of London, 2014.

Lengwiler, Martin. "Technologies of Trust: Actuarial Theory, Insurance Sciences, and the Establishment of the Welfare State in Germany and Switzerland around 1900." *Information and Organization* 13, no. 2 (April 2003): 131–150. https://doi.org/10.1016 /S1471-7727(02)00024-6.

Levy, Geoffrey. "Why 20,000 People Walked Out of a Job." *Daily Mail*, August 1, 1992.

Liffen, John. "Epsom, Britain's First Public Automatic Telephone Exchange." *International Journal for the History of Engineering & Technology* 82, no. 2 (July 2012): 210–232.

Liffen, John. "Telegraphy and Telephones." *Industrial Archaeology Review* 35, no. 1 (2013): 22–39.

Light, Jennifer S. "When Computers Were Women." *Technology and Culture* 40, no. 3 (1999): 455–483.

Light, Jennifer S. *From Warfare to Welfare: Defense Intellectuals and Urban Problems in Cold War America.* Baltimore, MD: Johns Hopkins University Press, 2003.

Lipartito, Kenneth. "When Women Were Switches: Technology, Work, and Gender in the Telephone Industry, 1890–1920." *American Historical Review* 99, no. 4 (1994): 1075–1111. https://doi.org/10.2307/2168770.

Lipartito, Kenneth. "Picturephone and the Information Age: The Social Meaning of Failure." *Technology and Culture* 44, no. 1 (2003): 50–81.

Lipartito, Kenneth. "Regulation Reconsidered: The Telephone Industry since 1975." *Entreprises et Histoire* 61, no. 4 (2010): 164–191.

Llewellyn, Mark. "Producing and Experiencing Harlow: Neighbourhood Units and Narratives of New Town Life 1947–53." *Planning Perspectives* 19, no. 2 (2004): 155–174.

"London's New Landmark." *Eagle and Swift*, May 23, 1964.

"Long Range Intelligence Division, Post Office Telecommunications Headquarters: Information Scientist." *New Scientist*, February 28, 1974.

"Long-Distance Transmission by Waveguide." *Post Office Electrical Engineers' Journal* 52, no. 3 (1959): 213–214.

"Longest Lightlines for UK Cities." *BT Technical Review*, Autumn 1982.

Lovell, Bernard. "The Challenge of Space Research." *Nature*, no. 4845 (1962): 195.

Lowe, Rodney. "Milestone or Millstone? The 1959–1961 Plowden Committee and Its Impact on British Welfare Policy." *Historical Journal* 40, no. 2 (1997): 463–491.

Lyon, David. *The Electronic Eye: The Rise of Surveillance Society*. Cambridge, UK: Polity Press, 1994.

MacDougall, Robert. *The People's Network: The Political Economy of the Telephone in the Gilded Age*. Philadelphia: University of Pennsylvania Press, 2014.

MacKenzie, Donald. "Opening the Black Boxes of Global Finance." *Review of International Political Economy* 12, no. 4 (2005): 555–576. https://doi.org/10.1080/09692 290500240222.

MacKenzie, Donald. *Material Markets: How Economic Agents Are Constructed*. Oxford: Oxford University Press, 2008.

MacKenzie, Donald. "Mechanizing the Merc: The Chicago Mercantile Exchange and the Rise of High-Frequency Trading." *Technology and Culture* 56, no. 3 (2015): 646–675. https://doi.org/10.1353/tech.2015.0102.

MacKenzie, Donald. "A Material Political Economy: Automated Trading Desk and Price Prediction in High-Frequency Trading." *Social Studies of Science* 47, no. 2 (2017): 172–194. https://doi.org/10.1177/0306312716676900.

MacKenzie, Donald. "Material Signals: A Historical Sociology of High-Frequency Trading." *American Journal of Sociology* 123, no. 6 (2018): 1635–1683.

MacKenzie, Donald, Daniel Beunza, Yuval Millo, and Juan Pablo Pardo-Guerra. "Drilling through the Allegheny Mountains: Liquidity, Materiality and High-Frequency Trading." *Journal of Cultural Economy* 5, no. 3 (August 2012): 279–296. https://doi.org /10.1080/17530350.2012.674963.

Mailland, Julien, and Kevin Driscoll. *Minitel: Welcome to the Internet*. Cambridge, MA: MIT Press, 2017.

"Main Event—Martlesham Heath: 100 Groundbreaking Years." Accessed July 21, 2017. https://mh100.org.uk/the-main-event/.

Malfona, Lina. "Building Silicon Valley. Corporate Architecture, Information Technology and Mass Culture in the Digital Age." *Histories of Postwar Architecture*, no. 4 (2019): 75–97. https://doi.org/10.6092/issn.2611-0075/9662.

Marchand, Roland. *Creating the Corporate Soul: The Rise of Public Relations and Corporate Imagery in American Big Business*. Berkeley: University of California Press, 1998.

Margot Lee Shetterly. *Hidden Figures: The Untold Story of the African American Women Who Helped Win the Space Race*. London: William Collins, 2016.

Martin, Michèle. "Communication and Social Forms: The Development of the Telephone, 1876–1920." *Antipode* 23, no. 3 (1991): 307–333. https://doi.org/10.1111/j .1467-8330.1991.tb00666.x.

Martin, Michèle. *"Hello Central?": Gender, Technology, and Culture in the Formation of Telephone Systems*. Montreal: McGill-Queen's University Press, 1991.

"Martlesham Heath: 100 Groundbreaking Years." Accessed July 20, 2017. https:// mh100.org.uk/.

"Martlesham Heath Township Proposals for about 10,000 People." *Surveyor and Municipal Engineer*, March 20, 1965, pp. 31–35.

Mayr, Otto. *Authority, Liberty, and Automatic Machinery in Early Modern Europe*. Baltimore, MD: Johns Hopkins University Press, 1986.

McCray, W. Patrick. *The Visioneers: How a Group of Elite Scientists Pursued Space Colonies, Nanotechnologies, and a Limitless Future*. Princeton, NJ: Princeton University Press, 2012.

McDonald, David A. "To Corporatize or Not to Corporatize (and If So, How?)." *Utilities Policy* 40 (2016): 107–114. https://doi.org/10.1016/j.jup.2016.01.002.

McGee, A. M. "Stating the Field: Institutions and Outcomes in Computer History." *IEEE Annals of the History of Computing* 34, no. 1 (2012): 104–103. https://doi.org/10.1109/MAHC.2012.14.

McGregor, Jay. "How Thatcher Killed the UK's Superfast Broadband before It Even Existed." *TechRadar*, March 12, 2014. http://www.techradar.com/news/world-of-tech/how-the-uk-lost-the-broadband-race-in-1990-1224784.

McGuire, Coreen Anne. "The Categorisation of Hearing Loss through Telephony in Inter-War Britain." *History and Technology* 35, no. 2 (2019): 138–155.

McKenna, Christopher D. *The World's Newest Profession: Management Consulting in the Twentieth Century*. Cambridge, UK: Cambridge University Press, 2006.

Mclaughlin, Charlotte. "Have Your Say on First 315 New Homes Being Built at Brightwell Lakes." *Ipswich Star*, June 29, 2021. https://www.ipswichstar.co.uk/news/housing/brightwell-lakes-housing-project-martlesham-consultation-8091938.

McLuhan, Marshall. *The Gutenberg Galaxy*. London: Routledge & Kegan Paul, 1962.

McNally, Kevin. *Corporate Venture Capital: Bridging the Equity Gap in the Small Business Sector*. London: Routledge, 1997.

McNeill, J. R., and Corinna R. Unger. "Introduction: The Big Picture." In *Environmental Histories of the Cold War*, edited by J. R. McNeill and Corinna R. Unger, 1–18. Cambridge, UK: Cambridge University Press, 2010.

McRae, Neil. "Labour Plans Broadband Communism!" Twitter, November 14, 2019. https://web.archive.org/web/20200227141748/https://twitter.com/neilmcrae/status/1195115144583557121.

Meades, Jonathan. *Museum without Walls*. London: Unbound Digital, 2012.

Meadows, Donella H., Dennis L. Meadows, Jørgen Randers, and William W. Behrens III. *The Limits to Growth*. New York: Universe Books, 1972.

Medina, Eden. *Cybernetic Revolutionaries: Technology and Politics in Allende's Chile*. Cambridge, MA: MIT Press, 2011.

Megginson, William L., and Jeffry M. Netter. "From State to Market: A Survey of Empirical Studies on Privatization." *Journal of Economic Literature* 39, no. 2 (2001): 321–389. https://doi.org/10.1257/jel.39.2.321.

Meller, Helen. *Towns, Plans and Society in Modern Britain*. Cambridge, UK: Cambridge University Press, 1997.

Merlo, D. "The Millimetric Waveguide System: The Current Situation." *Post Office Electrical Engineers' Journal* 69, no. 1 (1976): 34–37.

Merriman, J. H. H. "Men, Circuits and Systems in Telecommunications." *Post Office Electrical Engineers' Journal* 60, no. 4 (1968): 241–251.

Merriman, J. H. H., and D. W. G. Wass. "To What Extent Can Administration Be Mechanized?" In *Mechanisation of Thought Processes*. National Physical Laboratory. London: HMSO, 1959.

Merriman, Peter. *Driving Spaces: A Cultural-Historical Geography of England's M1 Motorway*. Malden, MA: Blackwell, 2007.

Messeri, Lisa, and Janet Vertesi. "The Greatest Missions Never Flown: Anticipatory Discourse and the 'Projectory' in Technological Communities." *Technology and Culture* 56, no. 1 (2015): 54–85.

Metro—A British Car to Beat the World. The Best of British Leyland. 1980. Reprint, London: BMIHT/David Weguelin Productions, 2005. https://shop.britishmotormuseum.co.uk /collections/dvds/products/the-best-of-british-leyland-hmfdvd-5004.

Michie, R. C. "Friend or Foe? Information Technology and the London Stock Exchange since 1700." *Journal of Historical Geography* 23, no. 3 (1997): 304–326.

Michie, Ronald C. *The City of London: Continuity and Change, 1850–1990*. Basingstoke, UK: Palgrave Macmillan, 1992. https://doi.org/10.1007/978-1-349-12322-3.

"Middlesex XI.14 (Includes: Willesden)." 1:2500. Southampton: Ordnance Survey, 1914. http://maps.nls.uk/view/103657949.

"Middlesex XI.14 (Includes: Willesden)." 1:2500. Southampton: Ordnance Survey, 1938. http://maps.nls.uk/view/103657946.

Midwinter, J. E. "Optical-Fibre Transmission Systems: Overview of Present Work." *Post Office Electrical Engineers' Journal* 70, no. 3 (1977): 146–153.

Millward, Robert. "European Governments and the Infrastructure Industries, c. 1840–1914." *European Review of Economic History* 8, no. 1 (2004): 3–28. https://doi .org/10.1017/S1361491604001030.

Millward, Robert. *Private and Public Enterprise in Europe: Energy, Telecommunications and Transport, 1830–1990*. Cambridge, UK: Cambridge University Press, 2005.

Millward, Robert. "Business and Government in Electricity Network Integration in Western Europe, c. 1900–1950." *Business History* 48, no. 4 (2006): 479–500. https:// doi.org/10.1080/00076790600808617.

Millward, Robert. "Geo-Politics versus Market Structure Interventions in Europe's Infrastructure Industries c. 1830–1939." *Business History* 53, no. 5 (2011): 673–687. https://doi.org/10.1080/00076791.2011.599595.

Milne, Graeme J. "British Business and the Telephone, 1878–1911." *Business History* 49, no. 2 (2007): 163–185.

Milne, Graeme J. "Business Districts, Office Culture and the First Generation of Telephone Use in Britain." *International Journal for the History of Engineering & Technology* 80, no. 2 (2010): 199–213.

Mirowski, Philip. *Machine Dreams: Economics Becomes a Cyborg Science*. Cambridge, UK: Cambridge University Press, 2008.

Mirowski, Philip, and Edward Nik-Khah. *The Knowledge We Have Lost in Information.* Oxford: Oxford University Press, 2017.

Mirowski, Philip, and Dieter Plehwe, eds. *The Road from Mont Pèlerin: The Making of the Neoliberal Thought Collective.* 2nd ed. Cambridge, MA: Harvard University Press, 2015.

Misa, Thomas J. "Understanding 'How Computing Has Changed the World.'" *IEEE Annals of the History of Computing* 29, no. 4 (2007): 52–63. https://doi.org/10.1109/MAHC.2007.4407445.

Misa, Thomas J., ed. *Gender Codes: Why Women Are Leaving Computing.* Hoboken, NJ: John Wiley & Sons, 2010.

Misa, Thomas J. *Digital State: The Story of Minnesota's Computing Industry.* Minneapolis: University of Minnesota Press, 2013.

Mitchell, Timothy. *Rule of Experts: Egypt, Techno-Politics, Modernity.* Berkeley: University of California Press, 2002.

Mitchell, Timothy. *Carbon Democracy: Political Power in the Age of Oil.* London: Verso, 2011.

"More Needles in Space." *New York Times*, May 8, 1963.

Moreton, Anthony. "Survey: Science Parks." *Financial Times*, January 21, 1983.

Morus, Iwan Rhys. "'The Nervous System of Britain': Space, Time and the Electric Telegraph in the Victorian Age." *British Journal for the History of Science* 33, no. 4 (2000): 455–475.

Mosca, Manuela. "On the Origins of the Concept of Natural Monopoly: Economies of Scale and Competition." *European Journal of the History of Economic Thought* 15, no. 2 (2008): 317–353. https://doi.org/10.1080/09672560802037623.

Mossop, Rebecca. "Infrastructures of Repair: Maintaining the Telephone System in Luxembourg." PhD, in progress.

Mounier-Kuhn, P. "On the History of the Data Processing Industry in France." *Engineering Science and Education Journal* 4, no. 1 (1995): 37–40.

Mounier-Kuhn, Pierre-E. "French Computer Manufacturers and the Component Industry, 1952–1972." *History and Technology* 11, no. 2 (1994): 195–216. https://doi.org/10.1080/07341519408581863.

Müller, Simone. *Wiring the World: The Social and Cultural Creation of Global Telegraph Networks.* New York: Columbia University Press, 2016.

Müller, Simone M., and Heidi J. S. Tworek. "'The Telegraph and the Bank': On the Interdependence of Global Communications and Capitalism, 1866–1914." *Journal of Global History* 10, no. 2 (2015): 259–283. https://doi.org/10.1017/S1740022815000066.

Muthesius, Stefan. *The Post-War University: Utopianist Campus and College.* New Haven, CT: Yale University Press, 2001.

Myddelton, D. R. "The British Approach to Privatisation." *Economic Affairs* 34, no. 2 (2014): 129–138. https://doi.org/10.1111/ecaf.12063.

Negrine, Ralph. "Cable Television in Great Britain." In *Cable Television and the Future of Broadcasting*, edited by Ralph Negrine, 103–133. London: Croom Helm, 1985.

"New Chief." *Post Office Telecommunications Journal*, Autumn 1972.

Newman, Karin. *The Selling of British Telecom*. London: Holt, Rinehart and Winston, 1986.

"No Adastral New Town." Accessed July 20, 2017. http://noadastralnewtown.com/.

Noam, Eli. *Telecommunications in Europe*. Oxford: Oxford University Press, 1992.

Noble, David F. *Forces of Production: A Social History of Industrial Automation*. Oxford: Oxford University Press, 1986.

Nuttall, Nick. "Well, Hello, How Nice to See You." *Times*, March 26, 1993.

Oakley, Brian, and Kenneth Owen. *Alvey: Britain's Strategic Computing Initiative*. Cambridge, MA: MIT Press, 1990.

Offer, Avner. "The Market Turn: From Social Democracy to Market Liberalism." *Economic History Review* 70, no. 4 (2017): 1051–1071.

O'Hara, Glen. *From Dreams to Disillusionment: Economic and Social Planning in 1960s Britain*. Basingstoke: Palgrave Macmillan, 2007.

O'Mara, Margaret. *The Code: Silicon Valley and the Remaking of America*. New York: Penguin, 2019.

O'Mara, Margaret Pugh. *Cities of Knowledge: Cold War Science and the Search for the Next Silicon Valley*. Princeton, NJ: Princeton University Press, 2005.

"On the Lightlines . . ." *British Telecom Journal* 3, no. 3 (Autumn 1982): 15.

"Optical Fibre Systems." *BT Technical News*, July 1979.

Oren, Tami, and Mark Blyth. "From Big Bang to Big Crash: The Early Origins of the UK's Finance-Led Growth Model and the Persistence of Bad Policy Ideas." *New Political Economy* 24, no. 5 (2019): 605–622. https://doi.org/10.1080/13563467.2018.1473355.

Ortolano, Guy. *Thatcher's Progress: From Social Democracy to Market Liberalism through an English New Town*. Cambridge, UK: Cambridge University Press, 2019.

Otis, Laura. "The Metaphoric Circuit: Organic and Technological Communication in the Nineteenth Century." *Journal of the History of Ideas* 63, no. 1 (2002): 105–128. https://doi.org/10.2307/3654260.

Oudshoorn, Nellie, and Trevor Pinch. *How Users Matter: The Co-Construction of Users and Technology*. Cambridge, MA: MIT Press, 2005.

Pardo-Guerra, Juan Pablo. *Automating Finance: Infrastructures, Engineers, and the Making of Electronic Markets*. Cambridge, UK: Cambridge University Press, 2019.

Parker, Christopher, and Gillian Darley. "Martlesham Heath Village." *Architects' Journal* 170, no. 36 (1979): 485–503.

Parker, David. *The Official History of Privatisation*. Vol. 1, *The Formative Years, 1970–1987*. 2 vols. London: Routledge, 2009.

Parker, David, and Colin Kirkpatrick. "Privatisation in Developing Countries: A Review of the Evidence and the Policy Lessons." *Journal of Development Studies* 41, no. 4 (2005): 513–541. https://doi.org/10.1080/00220380500092499.

Parker, Edwin B. "Information and Society." *Annual Review of Information Science and Technology* 8 (1973): 345–373.

Parker-Jervis, George. "Cable TV Firms Set to Break BT Monopoly." *Observer*, March 10, 1991.

Parks, Lisa. *Cultures in Orbit: Satellites and the Televisual*. Durham, NC: Duke University Press, 2005.

Parks, Lisa, and James Schwoch, eds. *Down to Earth: Satellite Technologies, Industries and Cultures*. New Brunswick, NJ: Rutgers University Press, 2012.

Peck, Jamie. *Constructions of Neoliberal Reason*. Oxford: Oxford University Press, 2013.

Peck, Jamie, and Adam Tickell. "Neoliberalizing Space." *Antipode* 34, no. 3 (2002): 380–404. https://doi.org/10.1111/1467-8330.00247.

Peck, Jamie, and Adam Tickell. "Conceptualizing Neoliberalism, Thinking Thatcherism." In *Contesting Neoliberalism: Urban Frontiers*, edited by Helga Leitner, Jamie Peck, and Eric Sheppard, 26–50. New York: The Guilford Press, 2007.

Pellizzoni, Luigi, and Marja Ylönen, eds. *Neoliberalism and Technoscience: Critical Assessments*. London: Routledge, 2016.

Peters, Benjamin. *How Not to Network a Nation: The Uneasy History of the Soviet Internet*. Cambridge, MA: MIT Press, 2016.

Phillips, Deborah. "Black Minority Ethnic Concentration, Segregation and Dispersal in Britain." *Urban Studies* 35, no. 10 (1998): 1681–1702. https://doi.org/10.1080/0042098984105.

"Phone Foresight." *Guardian*, October 7, 1971.

Pinch, Trevor. "Technology and Institutions: Living in a Material World." *Theory and Society* 37, no. 5 (2008): 461–483.

Pitt, Douglas C. *The Telecommunications Function in the British Post Office: A Case Study of Bureaucratic Adaptation*. Farnborough, UK: Saxon House, 1980.

Plaiss, Adam. "From Natural Monopoly to Public Utility: Technological Determinism and the Political Economy of Infrastructure in Progressive-Era America." *Technology and Culture* 57, no. 4 (2016): 806–830. https://doi.org/10.1353/tech.2016.0108.

Post Office. *Full Automation of the Telephone System*. Cmnd. 303. London: HMSO, 1957.

Post Office. *Reorganisation of the Post Office*. Cmnd. 3233. London: HMSO, 1967.

Post Office. *Report on Post Office Development and Finance*. Cmd. 9576. London: HMSO, 1955.

Post Office. *Telephone Policy: The Next Steps*. Cmnd. 436. London: HMSO, 1958.

Post Office. *Telephone Service and the Customer*. London: HMSO, 1959.

Post Office. *The Inland Telephone Service in an Expanding Economy*. Cmnd. 2211. London: HMSO, 1963.

"Post Office at Grips with Suppliers." *Daily Telegraph*, February 10, 1972.

"Post Office Democratic Pioneers." *Financial Times*, January 5, 1978.

"Post Office Enters the Computer Age, The." *Post Office Telecommunications Journal* 17, no. 1 (1965): 2–4.

"Post Office Equipment Row Brought into the Open." *Financial Times*, February 2, 1972.

Post Office Review Committee. *Report of the Post Office Review Committee*. Cmnd. 6850. London: HMSO, 1977.

Prasad, Monica. *The Politics of Free Markets: The Rise of Neoliberal Economic Policies in Britain, France, Germany, and the United States*. Chicago: University of Chicago Press, 2006.

Preda, Alex. "Socio-Technical Agency in Financial Markets: The Case of the Stock Ticker." *Social Studies of Science* 36, no. 5 (2006): 753–782. https://doi.org/10.1177/0306312706059543.

Preda, Alex. *Framing Finance*. Chicago: University of Chicago Press, 2009.

Prichard, David. "Village Values." *RIBA Journal* 95, no. 8 (1988): 42–45.

Probert, David. "Systems Dynamics Modelling within the British Telecommunications Business." *Dynamica* 8, no. 2 (1982): 69–81.

Probert, David. "The Development of a Long-Range Planning Model for the British Telecommunications Business: 'From Initiation to Implementation.'" *Journal of the Operational Research Society* 32, no. 8 (1981): 695–719.

"Professor Merriman Retires . . . and So Does Dollis Hill." *Post Office Telecommunications Journal* 28, no. 4 (1976): 30.

"Protests Continue Abroad." *New York Times*, October 22, 1961.

Queen Dials Edinburgh. YouTube video. British Pathé, 1958. https://www.youtube.com/watch?v=wfH0Xr1rIcY.

Ramanadham, V. V., ed. *Privatisation: A Global Perspective*. London: Routledge, 1993.

Rand, Lisa Ruth. "Orbital Decay: Space Junk and the Environmental History of Earth's Planetary Borderlands." PhD diss., University of Pennsylvania, 2016.

Randell, Brian. "Of Men and Machines." In *Colossus: The Secrets of Bletchley Park's Codebreaking Computers*, edited by B. Jack Copeland, 141–149. Oxford: Oxford University Press, 2006.

Rankin, Joy Lisi. *A People's History of Computing in the United States*. Cambridge, MA: Harvard University Press, 2018.

Raymond, Jack. "U.S. Gives Soviet Report on Search of Fishing Vessel." *New York Times*, February 28, 1959.

"Recent Major Events in the Evolution of British Telecom's Optical-Fibre Network." *British Telecommunications Engineering* 1, no. 3 (1982): 178–179.

Richards, John. "Phone Trouble Is Going to Get Worse." *Daily Mail*, June 17, 1966.

Roberts, Richard, and David Kynaston. *City State: A Contemporary History of the City and How Money Triumphed*. London: Profile Books, 2002.

Roeber, Joe. "In This Concluding Article, Joe Roeber Analyzes the Attitudes of Top Management to Their Consultants." *Times*, July 7, 1969.

Rollings, Neil. "Cracks in the Post-War Keynesian Settlement? The Role of Organised Business in Britain in the Rise of Neoliberalism before Margaret Thatcher." *Twentieth Century British History* 24, no. 4 (2013): 637–659.

Rollings, Neil. "Organised Business and the Rise of Neoliberalism: The Confederation of British Industry 1965–1990s." In *The Neoliberal Age? Britain since the 1970s*, edited by Aled Davies, Ben Jackson, and Florence Sutcliffe-Braithwaite, 279–298. London: UCL Press, 2021. https://doi.org/10.14324/111.9781787356856.

Rowbotham, T. R. "Plans for a British Trial of Fibre to the Home." *British Telecommunications Engineering* 8, no. 2 (1989): 78–82.

Rozwadowski, Helen M. "Ocean's Depths." *Environmental History* 15, no. 3 (2010): 520–525.

Rudge, Alan. "Why the DTI Is Out of Order on Telecoms." *Observer*, May 26, 1991.

Russell, Andrew L. "Histories of Networking vs. the History of the Internet." Presented at the SIGCIS Workshop, Copenhagen, October 7, 2012. https://arussell.org/papers/russell-SIGCIS-2012.pdf.

Russell, Andrew L. *Open Standards and the Digital Age: History, Ideology, and Networks*. Cambridge, UK: Cambridge University Press, 2014.

Russell, Andrew L., and Lee Vinsel. "After Innovation, Turn to Maintenance." *Technology and Culture* 59, no. 1 (2018): 1–25. https://doi.org/10.1353/tech.2018.0004.

Russell, Andrew L., and Valérie Schafer. "In the Shadow of ARPANET and Internet: Louis Pouzin and the Cyclades Network in the 1970s." *Technology and Culture* 55, no. 4 (2014): 880–907.

Rybczynski, Witold. "Behind the Façade of Prince Charles's Poundbury." *Architect*, December 3, 2013. http://www.architectmagazine.com/Design/behind-the-facade-of-prince-charless-poundbury_o.

"Ryland's Row." *Economist*, February 5, 1972.

Sagarena, Roberto Lint. "Building California's Past: Mission Revival Architecture and Regional Identity." *Journal of Urban History* 28, no. 4 (2002): 429–444. https://doi.org/10.1177/0096144202028004003.

Samuel, Raphael. *Theatres of Memory: Past and Present in Contemporary Culture*. London: Verso, 1994.

Saunders, Ben. "'Crisis? What Crisis?' Thatcherism and the Seventies." In *Making Thatcher's Britain*, edited by Ben Jackson and Robert Saunders, 25–42. Cambridge, UK: Cambridge University Press, 2012.

Sawhney, Harmeet. "Information Superhighway: Metaphors as Midwives." *Media, Culture & Society* 18, no. 2 (1996): 291–314. https://doi.org/10.1177/016344396018002007.

Saxenian, AnnaLee. *Regional Advantage: Culture and Competition in Silicon Valley and Route 128*. Cambridge, MA: Harvard University Press, 1994.

Schafer, Valérie, and Benjamin G. Thierry, eds. *Connecting Women: Women, Gender and ICT in Europe in the Nineteenth and Twentieth Century*. Cham, Switzerland: Springer, 2015.

Schiller, Herbert I. *Information and the Crisis Economy*. Norwood, NJ: Ablex, 1984.

Schlombs, Corinna. *Productivity Machines*. Cambridge, MA: MIT Press, 2019.

Schmidt, Suzanne K., and Raymund Werle. *Coordinating Technology: Studies in the International Standardization of Telecommunications*. Cambridge, MA: MIT Press, 1998.

Schot, Johan, and Vincent Lagendijk. "Technocratic Internationalism in the Interwar Years: Building Europe on Motorways and Electricity Networks." *Journal of Modern European History* 6, no. 2 (2008): 196–217. https://doi.org/10.17104/1611 -8944_2008_2_196.

Schumacher, E. F. *Small Is Beautiful: A Study of Economics As If People Mattered*. London: Blond and Briggs, 1973.

Schwoch, James. *Global TV: New Media and the Cold War, 1946–69*. Urbana: University of Illinois Press, 2008.

Schwoch, James. *Wired into Nature: The Telegraph and the North American Frontier*. Urbana: University of Illinois Press, 2018.

Scott, Peter. "Still a Niche Communications Medium: The Diffusion and Uses of the Telephone System in Interwar Britain." *Business History* 53, no. 6 (2011): 801–820.

Scranton, Philip. "None-Too-Porous Boundaries: Labor History and the History of Technology." *Technology and Culture* 29, no. 4 (1988): 722–743. https://doi.org/10 .2307/3105043.

"Screen Test." *Guardian*, February 26, 1982.

Seefried, Elke. "Towards the Limits to Growth? The Book and Its Reception in West Germany and Britain 1972–73." *German Historical Institute London Bulletin* 33, no. 1 (2011): 3–37.

Seely, Bruce Edsall. *Building the American Highway System: Engineers as Policy Makers*. Philadelphia: Temple University Press, 1987.

Shannon, Claude E. "A Mathematical Theory of Communication, Part 1." *Bell System Technical Journal* 27 (1948): 379–423.

Shannon, Claude E. "A Mathematical Theory of Communication, Part 2." *Bell System Technical Journal* 27 (1948): 623–656.

Sheppard, S. H. "The Leighton Buzzard Electronic Telephone Exchange." *Post Office Electrical Engineers' Journal* 59, no. 4 (1967): 255–261.

Skousen, Mark. *Vienna & Chicago, Friends or Foes?: A Tale of Two Schools of Free-Market Economics*. Washington, DC: Capital Press, 2005.

Slobodian, Quinn. *Globalists: The End of Empire and the Birth of Neoliberalism*. Cambridge, MA: Harvard University Press, 2018.

Slotten, Hugh Richard. "Satellite Communications, Globalization, and the Cold War." *Technology and Culture* 43, no. 2 (2002): 315–350.

Slotten, Hugh Richard. "International Governance, Organizational Standards, and the First Global Satellite Communication System." *Journal of Policy History* 27, no. 3 (2015): 521–549.

Slotten, Hugh Richard. "The International Telecommunications Union, Space Radio Communications, and U.S. Cold War Diplomacy, 1957–1963." *Diplomatic History* 37, no. 2 (2013): 313–371.

Smith, J. T. "So These Are 'Bits': Briefly." *Post Office Magazine*, April 1958.

Smith, Roger. "BT Bites Back . . . New Cable More Than a Match for 'Jaws.'" *BT Journal* 8, no. 2 (1987): 38–41.

Snoke, L. R. "Resistance of Organic Materials and Cable Structures to Marine Biological Attack." *Bell System Technical Journal* 36, no. 5 (1957): 1095–1127.

Starosielski, Nicole. *The Undersea Network*. Durham, NC: Duke University Press, 2015.

Stedman-Jones, Daniel. *Masters of the Universe: Hayek, Friedman and the Birth of Neoliberal Politics*. Princeton, NJ: Princeton University Press, 2014.

Stefik, Mark J., ed. *Internet Dreams: Archetypes, Myths, and Metaphors*. Cambridge, MA: MIT Press, 1996.

Stevens, Richard. "The Evolution of Privatisation as an Electoral Policy, c. 1970–90." *Contemporary British History* 18, no. 2 (2004): 47–75.

Stine, Jeffrey K., and Joel A. Tarr. "At the Intersection of Histories: Technology and the Environment." *Technology and Culture* 39, no. 4 (1998): 601–640.

Sumner, James. "Defiance to Compliance: Visions of the Computer in Postwar Britain." *History and Technology* 30, no. 4 (2014): 309–333.

"'Super' Link Installed." *British Telecom Journal* 6, no. 4 (1985): 41.

Sutcliffe-Braithwaite, Florence, Aled Davies, and Ben Jackson. "Introduction: A Neoliberal Age?" In *The Neoliberal Age? Britain since the 1970s*, edited by Aled Davies, Ben Jackson, and Florence Sutcliffe-Braithwaite, 1–29. London: UCL Press, 2021. https://doi.org/10.14324/111.9781787356856.

Sweetman, Ashley. *Cyber and the City: Securing London's Banks in the Computer Age*. Cham, Switzerland: Springer, 2022.

"Switchboard In Orbit." *TV Guide*, December 4, 1971.

"System X Arrives Early." *Telecom Today*, May 1980.

Telecommunication Services for the 1990s. YouTube video. The Post Office, 1969. https://www.youtube.com/watch?v=EUcF_OuV19k.

"Telecommunications Services for the 1990s." *The Archiveologists*. BBC, February 27, 2018.

"Telecom's Optical Triumph." *BT Management News*, March 1982.

"Telephone Delays Up." *Times*, February 20, 1967.

Temin, Peter, and Louis Galambos. *The Fall of the Bell System: A Study in Prices and Politics*. Cambridge, UK: Cambridge University Press, 1987.

Thatcher, Mark. *The Politics of Telecommunications: National Institutions, Convergence, and Change in Britain and France*. Oxford: Oxford University Press, 1999.

Thatcher, Mark. *Internationalisation and Economic Institutions: Comparing the European Experience*. Oxford: Oxford University Press, 2007.

Thomas, Graham, and Sally Wyatt. "Shaping Cyberspace—Interpreting and Transforming the Internet." *Research Policy* 28, no. 7 (1999): 681–698.

Thomas, William. *Rational Action: The Sciences of Policy in Britain and America, 1940–1960*. Cambridge, MA: MIT Press, 2015.

Thomas, William, and Lambert Williams. "The Epistemologies of Non-Forecasting Simulations, Part I: Industrial Dynamics and Management Pedagogy at MIT." *Science in Context* 22, no. 2 (2009): 245–270.

Thompson, Helen. "The Thatcherite Economic Legacy." In *The Legacy of Thatcherism: Assessing and Exploring Thatcherite Social and Economic Policies*, edited by Stephen Farrall and Colin Hay, 33–68. Oxford: Oxford University Press, 2014.

Thring, M. W., and E. R. Laithwaite. *How to Invent*. London: Macmillan, 1977.

"Tommy Flowers Institute: BT Adastral Park." Accessed July 21, 2017. http://atadastral.co.uk/about/tommy-flowers-institute/.

Tully, John. "A Victorian Ecological Disaster: Imperialism, the Telegraph, and Gutta-Percha." *Journal of World History* 20, no. 4 (2009): 559–579.

Tuomi, Ilkka. *Networks of Innovation: Change and Meaning in the Age of the Internet*. Oxford: Oxford University Press, 2002.

Turnbull, Thomas. "Simulating the Global Environment: The British Government's Response to the Limits to Growth." In *Histories of Technology, the Environment and Modern Britain*, edited by Jon Agar and Jacob Ward, 271–299. London: UCL Press, 2018.

Turner, Fred. *From Counterculture to Cyberculture: Stewart Brand, the Whole Earth Network, and the Rise of Digital Utopianism*. Chicago: University of Chicago Press, 2006.

Turner, Rachel S. *Neo-Liberal Ideology: History, Concepts and Policies*. Edinburgh: Edinburgh University Press, 2008.

"Twentieth Century Village." *Building Design*, no. 406 (1978): 8.

Tworek, Heidi J. S. "How Not to Build a World Wireless Network: German-British Rivalry and Visions of Global Communications in the Early Twentieth Century." *History and Technology* 32, no. 2 (2016): 178–200. https://doi.org/10.1080/07341512.2016.1217599.

Tworek, Heidi J. S. *News from Germany: The Competition to Control World Communications, 1900–1945*. Cambridge, MA: Harvard University Press, 2019.

Tynan, Nicola. "Mill and Senior on London's Water Supply: Agency, Increasing Returns, and Natural Monopoly." *Journal of the History of Economic Thought* 29, no. 1 (2007): 49–65. https://doi.org/10.1080/10427710601178302.

"UK Ministers Defend Chinese Deals after Security Risk Warning." *BBC News*, June 6, 2013. http://www.bbc.co.uk/news/uk-politics-22795226.

"Union Approves the 'Friendly Telephone.'" *Manchester Guardian*, May 16, 1959.

University of Essex. "University of Essex Celebrates Four Decades of Links with BT." *University of Essex News*, May 28, 2021. https://www.essex.ac.uk/news/2021/05/28/celebrating-links-between-essex-and-bt.

"University of Essex: Chair in Telecommunication Systems." *Observer*, January 15, 1967.

"University of Essex: Electronic Engineering." *Guardian*, December 22, 1967.

"University of Essex/Post Office Research Centre. SRC Case Studentship: Optical Communications." *Guardian*, May 29, 1979.

Usselman, Steven W. *Regulating Railroad Innovation: Business, Technology, and Politics in America, 1840–1920*. Cambridge, UK: Cambridge University Press, 2002.

Usselman, Steven W. "Unbundling IBM: Antitrust and the Incentives to Innovation in American Computing." In *The Challenge of Remaining Innovative: Insights from Twentieth-Century American Business*, edited by Sally H. Clarke, Naomi R. Lamoreaux, and Steven W. Usselman, 249–280. Stanford, CA: Stanford University Press, 2009.

Vallis, Peter. "Martlesham Heath Village." *ERA*, no. 45 (August 1977): 64–68.

"Van Allen Sees Science 'Clique.'" *New York Times*, December 31, 1962.

Veljanovski, Cento G. *Selling the State: Privatisation in Britain*. London: Weidenfeld and Nicolson, 1987.

Venugopal, Rajesh. "Neoliberalism as Concept." *Economy and Society* 44, no. 2 (2015): 165–187. https://doi.org/10.1080/03085147.2015.1013356.

Vernon, James. "The Local, the Imperial and the Global: Repositioning Twentieth-Century Britain and the Brief Life of Its Social Democracy." *Twentieth Century British History* 21, no. 3 (2010): 404–418.

Vieille Blanchard, Elodie. "Modelling the Future: An Overview of the 'Limits to Growth' Debate." *Centaurus* 52, no. 2 (2010): 91–116.

Vinen, Richard. *Thatcher's Britain: The Politics and Social Upheaval of the 1980s*. London: Simon & Schuster, 2010.

Vinsel, Lee, and Andrew L. Russell. *The Innovation Delusion: How Our Obsession with the New Has Disrupted the Work That Matters Most*. New York: Currency, 2020.

Waff, Craig B. "Project Echo, Goldstone, and Holmdel: Satellite Communications as Viewed from the Ground Station." In *Beyond the Ionosphere: Fifty Years of Satellite Communication*, edited by Andrew J. Butrica, 41–50. Washington, DC: NASA History Office, 1997.

Wakeman, Rosemary. *Practicing Utopia: An Intellectual History of the New Town Movement*. Chicago: University of Chicago Press, 2016.

Ward, Jacob. "Oceanscapes and Spacescapes in North Atlantic Communications." In *Histories of Technology, the Environment and Modern Britain*, edited by Jon Agar and Jacob Ward, 186–205. London: UCL Press, 2018. http://doi.org/10.2307/j.ctvqhsmr.14.

Ward, Jacob. "Computer Models and Thatcherist Futures: From Monopolies to Markets in British Telecommunications." *Technology and Culture* 61, no. 3 (July 2020): 843–870.

Ward, Jacob. "Nineteen Eighty-Four in the British Telephone System: Computers, Science Fiction, and Thatcherism in British Telecom." In *Futures*, edited by Jenny Andersson and Sandra Kemp, 174–190. Oxford: Oxford University Press, 2021. 10.1093/oxfordhb/9780198806820.013.11.

Ward, K. E. "The Heart That Makes the Tower Tick." *Post Office Telecommunications Journal* 20, no. 2 (1968): 22–27.

Ward, Kevin, and Kim England. "Introduction: Reading Neoliberalism." In *Neoliberalization: States, Networks, People*, edited by Kim England and Kevin Ward, 1–22. Malden, MA: Blackwell, 2007.

Ward, Stephen V. "Consortium Developments Ltd and the Failure of 'New Country Towns' in Mrs Thatcher's Britain." *Planning Perspectives* 20, no. 3 (2005): 329–359. https://doi.org/10.1080/02665430500130290.

Webster, Frank. *Theories of the Information Society*. London: Routledge, 2014.

Weppler, H. Edward. "The Radio Spectrum Squeeze." *Bell Telephone Magazine* 47, no. 6 (1968): 28–32.

West, W. "Facilities for Experimental Work in the Engineering Department." *Post Office Electrical Engineers' Journal* 47, no. 1 (1954): 10–14.

Westin, Alan F. *Privacy and Freedom*. London: Bodley Head, 1967.

Wetherell, Sam. "Freedom Planned: Enterprise Zones and Urban Non-Planning in Post-War Britain." *Twentieth Century British History* 27, no. 2 (2016): 266–289. https://doi.org/10.1093/tcbh/hww004.

White, R. W. "Waveguides—Highways of Communication." *Post Office Telecommunications Journal* 22, no. 3 (1970): 5.

Whitfield, Martin, and Mary Fagan. "BT Operators to Hold Strike Vote over Job Losses." *Independent*, September 3, 1991.

Whyte, J. S. "Telecommunications." *Guardian*, September 17, 1970.

Whyte, William H. *The Organization Man*. Garden City, NY: Doubleday, 1956.

Wiener, Norbert. *Cybernetics: Or Control and Communication in the Animal and Machine*. New York: John Wiley & Sons, 1948.

Wilby, Dave. "Inside Adastral: BT's Belgium-Sized Broadband Boffinry Base." March 26, 2013. https://www.theregister.co.uk/2013/03/26/geeks_guide_adastral_park/.

Williamson, Adrian. "The Bullock Report on Industrial Democracy and the Post-War Consensus." *Contemporary British History* 30, no. 1 (2016): 119–149.

Wilson, Harold. "Labour's Plan for Science." Presented at the Labour Party Annual Conference, Scarborough, October 1, 1963.

Wincott, Daniel. "Thatcher: Ideological or Pragmatic?" *Contemporary Record* 4, no. 2 (1990): 26–28. https://doi.org/10.1080/13619469008581120.

Winner, Langdon. "Mythinformation in the High-Tech Era." *Bulletin of Science, Technology & Society* 4, no. 6 (1984): 582–596.

Wise, Deborah. "Warning Bells Ring for BT and Mercury." *Guardian*, October 11, 1990.

Wise, M. Norton, ed. *The Values of Precision*. Princeton, NJ: Princeton University Press, 1995.

Wójcik, Dariusz. *The Global Stock Market: Issuers, Investors, and Intermediaries in an Uneven World*. Oxford: Oxford University Press, 2011.

"World's First Commercial 140 Mbit/s Optical Link Fibre Tested." *BT Technical News*, Winter 1983.

"World's First PCM Exchange, The." *Post Office Telecommunications Journal* 20, no. 2 (1968): 7–11.

Wright, Nigel. "The Formulation of British and European Policy toward an International Satellite Telecommunications System: The Role of the British Foreign Office." In *Beyond the Ionosphere: Fifty Years of Satellite Communication*, edited by Andrew J. Butrica, 157–170. Washington, DC: NASA History Office, 1997.

Wright, Patrick. *On Living in an Old Country: The National Past in Contemporary Britain*. London: Verso, 1985.

Wyatt, Sally. "Danger! Metaphors at Work in Economics, Geophysiology, and the Internet." *Science, Technology, & Human Values* 29, no. 2 (2004): 242–261.

Yates, David M. *Turing's Legacy: A History of Computing at the National Physical Laboratory 1945–1995*. London: Science Museum, 1997.

Yates, JoAnne. *Control through Communication: The Rise of System in American Management*. Baltimore, MD: Johns Hopkins University Press, 1993.

Yates, JoAnne, and Craig N. Murphy. *Engineering Rules: Global Standard Setting since 1880*. Baltimore, MD: Johns Hopkins University Press, 2019.

Yost, Jeffrey R. *Making IT Work: A History of the Computer Services Industry*. Cambridge, MA: MIT Press, 2017.

Yueh, Linda. "Huawei Boss Says US Ban 'Not Very Important.'" *BBC News*, October 16, 2014. http://www.bbc.co.uk/news/business-29620442.

INDEX

ACE. *See* Automatic Computing Engine

Adastral Park, 149–152, 164–170, 224. *See also* Martlesham Heath

ADMITS (Adaptable Dispersed Modular Integrated Telecommunications System), 92, 112. *See also* Digital vision; Integrated digital network; System X

Advisory Committee on Science and Technology, 140, 145

Advisory Group on Systems Definition, 58, 92, 100–101, 132. *See also* System X

AGSD. *See* Advisory Group on Systems Definition

Air conditioning. *See* Laboratory design

ALEM (A Local Exchange Model), 80, 87, 111, 223. *See also* Simulations; TXE4
development of, 60–61
dispute over, 93–99

ALF (Automatic Letter Facer), 90. *See also* Robots

Alvey Programme, 212

Andover Earth Station, Maine, 179, 183, 191. *See also* Earth stations

Annan Committee on the future of broadcasting, 135–136. *See also* Cable television; Television broadcasting;

Artificial lighting. *See* Laboratory design

Association of Telecommunications Users, 208, 219. *See also* Business users; Financial users; Users

AT&T, 224–225
at Adastral Park, 164–165, 170
advertising of international services by, 171–172, 195–196
domestic monopoly of, 9, 13, 172, 193, 196, 198
and INTELSAT, 193–194
metaphors used by, 116, 128–129
and regulation of transatlantic communications, 173, 178–179, 196–198 (*see also* FCC)
and submarine cable maintenance, 191–192 (*see also* North Atlantic Cable Maintenance Agreement; North Atlantic Systems Conference)
and TAT-1, 174–179 (*see also* TAT-1)
and TAT-8, 193–194 (*see also* TAT-8)
and telephone operators, 87, 91

Arthur L. Norberg
Computers and Commerce: A Study of Technology and Management at Eckert-Mauchly Computer Company, Engineering Research Associates, and Remington Rand, 1946–1957

Emerson W. Pugh
Building IBM: Shaping an Industry and Its Technology

Emerson W. Pugh
Memories That Shaped an Industry

Emerson W. Pugh, Lyle R. Johnson, and John H. Palmer
IBM's Early Computers: A Technical History

Kent C. Redmond and Thomas M. Smith
From Whirlwind to MITRE: The R&D Story of the SAGE Air Defense Computer

Alex Roland with Philip Shiman
Strategic Computing: DARPA and the Quest for Machine Intelligence, 1983–1993

Raúl Rojas and Ulf Hashagen, editors
The First Computers—History and Architectures

Corinna Schlombs
Productivity Machines: German Appropriations of American Technology from Mass Production to Computer Automation

Dinesh C. Sharma
The Outsourcer: A Comprehensive History of India's IT Revolution

Dorothy Stein
Ada: A Life and a Legacy

Christopher Tozzi
For Fun and Profit: A History of the Free and Open Source Software Revolution

John Vardalas
The Computer Revolution in Canada: Building National Technological Competence, 1945–1980

Maurice V. Wilkes
Memoirs of a Computer Pioneer

Jeffrey R. Yost
Making IT Work: A History of the Computer Services Industry

Thomas Haigh and Paul E. Ceruzzi
A New History of Modern Computing

Daniel D. Garcia-Swartz and Martin Campbell-Kelly
Cellular: An Economic and Business History of the International Mobile-Phone Industry

Victor Petrov
Balkan Cyberia: Bulgarian Modernization, Computers, and the World, 1963–1989

Jacob Ward
Visions of a Digital Nation: Market and Monopoly in British Telecommunications